AF478472

RETAINING WALLS
Study of Passive Resistance in Foundation Structures

Other volumes in the
Series on Rock and Soil Mechanics

W. Reisner, M. v. Eisenhart Rothe:
Bins and Bunkers for Handling Bulk Materials
— Practical Design and Techniques —
1971

W. Dreyer:
The Science of Rock Mechanics
Part I: Strength Properties of Rocks
1972

T. H. Hanna:
Foundation Instrumentation
1973

C. E. Gregory:
Explosives for North American Engineers
1973

M. & A. Reimbert:
Retaining Walls Vol. I
— Anchorages and Sheet Piling —
1974

Vutukuri, Lama, Saluja:
Handbook on Mechanical Properties
of Rocks Vol. I
1974

Editor-in-Chief
Professor Dr. H. Wöhlbier

Series on Rock and Soil Mechanics
Vol. 2 (1974/76) No. 2

RETAINING WALLS

Volume II

Study of Passive Resistance
in Foundation Structures

by

Marcel L. Reimbert
Consulting Engineer

and

Andrew M. Reimbert
Consulting Engineer
Member, Committee „SILOS" of the American Concrete Institute

First Edition
1976

TRANS TECH PUBLICATIONS

Distributed
in the USA and Canada by
TRANS TECH PUBLICATIONS
411 Long Beach Pkwy.
Bay Village, Ohio 44140
USA

and worldwide by
TRANS TECH S. A.
CH-4711 Aedermannsdorf
Switzerland

Copyright © 1976 by
Trans Tech Publications
Clausthal, Germany

Translated by C. V. Amerongen, M. Sc., M.I.C.E.
for OLYMPIA TRANSLATION SERVICE, London

International Standard Book Number
ISBN 0-87849-013-2

Library of Congress Catalog Card Number
LCC 74-77789

Printed in Germany

1. Preface

Volume I, entitled "Retaining Walls, Anchorages and Sheet Piling", showed how the values of the **maximum thrust** and of the **minimum passive** resistance exerted upon, or developed by, retaining walls in general could be calculated.

These optimum values correspond to the **"at rest"** state of equilibrium of the soil, this being **the only state compatible with proper stability of retaining structures.**

Nevertheless, in special cases where such structures undergo displacements whereby they push back the retained soil upon itself, it used to be known merely that the magnitude of the passive resistance reaction developed by that soil would increase rapidly and considerably, though the relationship between the displacements and the stresses resulting therefrom had never been clearly established.

The extremely accurate experiments reported in Volume II have revealed the **deformation relationship** for a granular material under loading whereby it develops **translatory passive resistance** or **rotational passive resistance.** This deformation is defined by a **linear function of the passive resistance coefficient** in the elasto-plastic stage of equilibrium of the material.

The special problem presented by the equilibrium of soil developing passive resistance is therefore unambiguously solved, for it is henceforth possible to know the displacements that a retaining structure would undergo, depending on the value of the passive resistance coefficient corresponding to the equilibrium of the retained material tending to be pushed back upon itself by forces of determinate magnitude.

The information yielded by the researches reported in this book is thus directly applicable to the design of retaining walls, sheet piling, diaphragm walls, etc.

The information presented in this book embodies the results of our personal and unbiased efforts. It necessitated upwards of a thousand experiments of various kinds, performed with equipment designed entirely by ourselves and without any financial assistance from outside sources.

Nevertheless, despite the effort and expense involved, we carried out a properly planned experimental program that enabled us to ascertain and interpret the passive resistance phenomena in a most exhaustive manner.

Thus we have devoted, separate chapters to the effect of scale in models, to dilatancy, to the effect of displacement of the retaining wall in its own plane, to the surface condition of the wall, etc.

As regards this last-mentioned point, we knowingly confined ourselves to what was apparently a small displacement as soon as it was found to be sufficient to reveal that the surface condition of the wall was of no influence on the passive resistance at rest, while at the same time confirming the validity of the linear relationship between the deformations and the forces which has been proposed by us with regard to the phenomena of passive resistance.

This preliminary notice will thus have served to inform our readers of the scope and conditions of the present treatise which carries these researches to their proper conclusion in providing the interpretation of the phenomena of translatory and of rotational passive resistance.

Paris,   Marcel L. Reimbert
January 1976   Andrew M. Reimbert

Contents

2. Introduction

The experiments forming the subject of the present Volume II were carried out with the aid of procedures significantly different from those employed in the experimental work concerned with the measurement of thrust and passive resistance "at rest", as reported in Volume I.

The reader's attention is more particularly directed to the fact that the experiments relating to passive resistance with displacement of the retaining wall (as represented by a diaphragm in the investigations) provide rigorous confirmation of the previously proposed formulas for defining the state of equilibrium at rest of granular materials having a horizontal or an inclined top surface, with or without surcharge.

The information yielded by these researches as a whole thus thoroughly covers the whole range of the subject and provides a complete cross-check of all the formulas.

3. Granular Materials Used

Already in the course of the first experiments undertaken for the preliminary investigation of the passive resistance phenomena that were to be studied it became evident that tests performed with only one granular material would be insufficient to provide full information. We therefore used granular materials of different kinds and differing greatly from one another in their mechanical properties, more particularly as regards the density γ and the minimum angle of internal friction φ_0 which can be taken as equal to the angle of repose α.

Thus the definitive experiments reported in this volume were carried out with the following five cohesionless granular materials, with bulk densities ranging over a ratio of about 2 : 1. Their minimum angles of internal friction differ by more than 25% (as in the case of millet compared with Var sand) or are practically identical despite considerable difference in density (as in the case of hard wheat and Seine sand)*.

3.1. Cereals

Millet:	$\gamma = 730 \text{ kg/m}^3$	$\varphi_0 \cong \alpha = 28° \ 30'$
Hard wheat:	$\gamma = 790 \text{ kg/m}^3$	$\varphi_0 \cong \alpha = 31° \ 30'$

3.2. Sands

Fine sand (from the beach at Garoupe-Antibes):

$$\gamma = 1\ 380 \text{ kg/m} \qquad \varphi_0 \cong \alpha = 33° \ 40'$$

Crushed stone sand from the Var region of France:

$$\gamma = 1\ 420 \text{ kg/m}^3 \qquad \varphi_0 \cong \alpha = 36° \ 30'$$

Sand from the river Seine:

$$\gamma = 1\ 550 \text{ kg/m}^3 \qquad \varphi_0 \cong \alpha = 34° \ 40'$$

---

* For symbols used, see Section 4.1.

Figures 1 to 5 show photographs of the materials used.

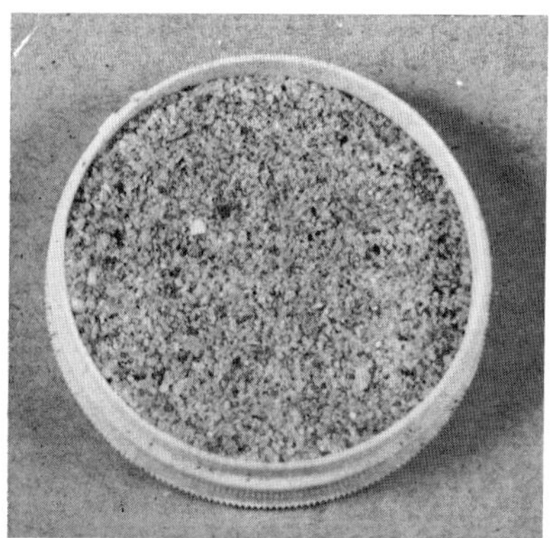

Fig. 1: Fine sand.

Fig. 2: Var sand.

Fig. 3: Seine sand.

Fig. 4: Millet.

Fig. 5: Wheat.

Figures 6 to 8 show grading curves for the sands, according to the report
of the C.E.B.T.P., 1973.

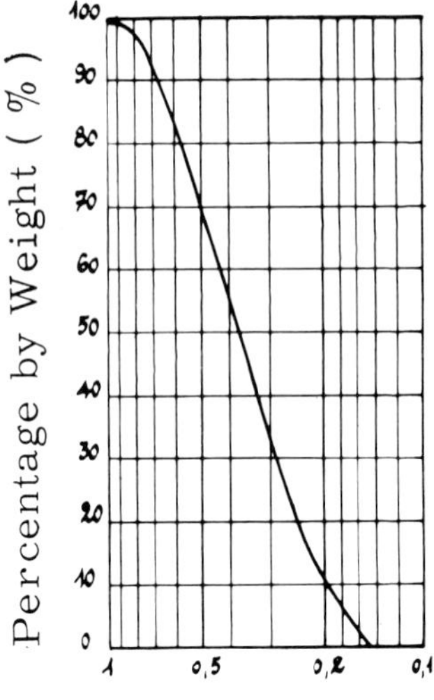

Fig. 6: Fine sand from Garoupe.

Fig. 7: Seine sand.

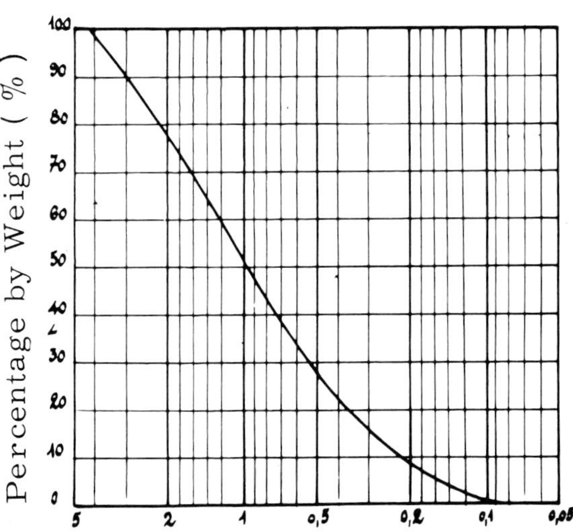

Fig. 8: Var crushed stone sand.

4. Deformations Associated with Passive Resistance

4.1. List of Symbols

Granular Material

γ: bulk density

α: angle of repose
- $-\alpha$ below the horizontal plane
- $+\alpha$ above the horizontal plane

α':   angle of inclination of the top surface of the material
- $-\alpha'$ below the horizontal plane
- $+\alpha'$ above the horizontal plane

φ: angle of internal friction of the material

φ_0:   minimum angle of internal friction

φ' or θ: angle of friction of the material in contact with the rear (retaining) face of the wall

Forces

F: force required to push back the material

p: active pressure (per unit area)

P: thrust (resultant force due to active pressure)

b: passive pressure (per unit area)

B: passive resistance (resultant force due to passive pressure)

Relative minimum **translatory** passive resistance for zero displacement:

B_0:   passive resistance in the case of material with a horizontal top surface

B_0':   passive resistance in the case of material with a top surface inclined at an angle $+\alpha'$

B_0'':   passive resistance in the case of material with a top surface inclined at an angle $-\alpha'$

Relative minimum **rotational** passive resistance for zero displacement:

$B_{0.r}$:   passive resistance in the case of material with a horizontal top surface

$B'_{0.r}$:   passive resistance in the case of material with a top surface inclined at an angle $+\alpha'$

$B''_{0.r}$:   passive resistance in the case of material with a top surface inclined at an angle $-\alpha'$

$B_{0.cb}$:   relative minimum **toe resistance** for zero displacement

Coefficients

K_0:   thrust coefficient "at rest"

K_a: active thrust coefficient

K_p: passive resistance coefficient in general

K_{p1} or k_1: translatory passive resistance coefficient

K_{p2} or k_2: rotational passive resistance coefficient

K_{p3} or k_3: toe resistance coefficient

k_p: passive resistance coefficient equal to the ratio of the force F (required to push back the material) and the relative minimum passive resistance, for zero displacement, in the case of translatory or rotational or toe resistance considered

4.2. Experimental Equipment

Figures 9, 10 and 11 illustrate the experimental equipment and system used. A vertical diaphragm (*e*) is fixed to a mobile metal carriage (*m*) which rests freely on a flat base through the interposition of three large accurately sized polished-steel bearing balls placed on a glass plate so as to eliminate all rolling friction (Fig. 9).

Fig. 9: Fixed screens of Fig. 13 not shown.

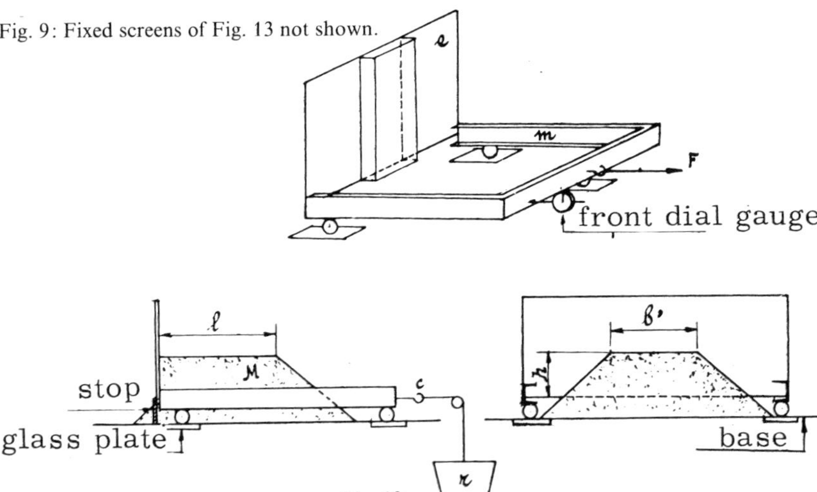

A pull (tensile force) F can be applied to the carriage so that it then moves on its three ball bearings. This force is exerted by a weighted receptacle (r) attached to a string passing round a pulley and attached to a hook (c) on the carriage (Fig. 10). First, the rolling system described above is calibrated by the application of a force F_0 which equilibrates the inertial force to be overcome in order to set the carriage in motion. Next, the granular material (m) is placed, while the carriage is arrested by a stop preventing its rearward movement when the material exerts its thrust. Then known quantities of water are poured into the receptacle (r) and the displacements of the diaphragm are read – to within a micron accuracy – on a dial gauge.

A second dial gauge installed at the rear of the diaphragm, in contact with it, enables the displacement measurements obtained with the first gauge to be checked.

Fig. 11: Test Apparatus.

4.3. Configuration of the Granular Material in the Experiments

The shape and dimensions of the mass of material retained by the diaphragm in the experimental set-up was the subject of a systematic investigation, as described in Section 13, as a result of which we decided to carry out all our definitive tests with a mass having a trapezoidal cross-section of height h' and having a horizontal top surface of width b' and length l (Figs. 10, 11 and 12).

As the diaphragm is raised some distance above the base, the height h of the part of the mass in contact with the diaphragm is less than h'. The difference between these two heights h and h' was varied by an amount ranging from 2 cm to 5 cm in order to observe whether there might be some kind of bottom effect to be taken into account. In the event, no such effect was found to exist (Fig. 12).

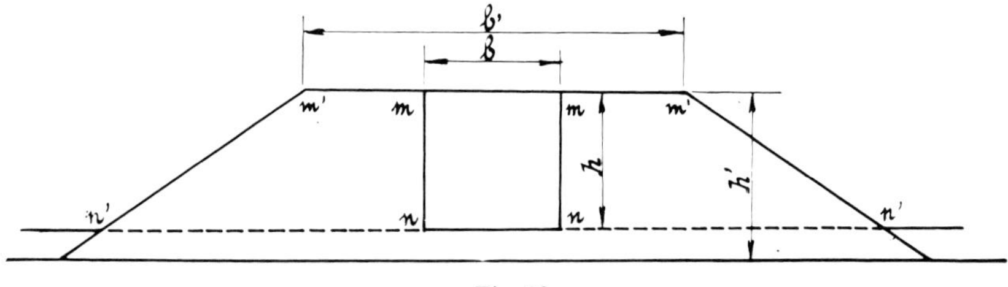

Fig. 12

The height h was varied from 10 to 12, 15, 20, 25 and 30 cm, and the width b' was at first – for lack of adequate information on the scale effect of the models – varied from 3 to 5 times the width b of the diaphragm. Finally, b' was reduced to 1.2 b after the phenomenon of dilatancy had more particularly been investigated (see Section 12).

The trapezium-shaped parts mm' nn' of the granular mass were screened by fixed diaphragms (Fig. 13) so that the tensile forces exerted on the movable carriage would be transmitted to the granular material by the rectangular part of the diaphragm (situated between the two screened parts) acting as a "piston" of height h and width b.

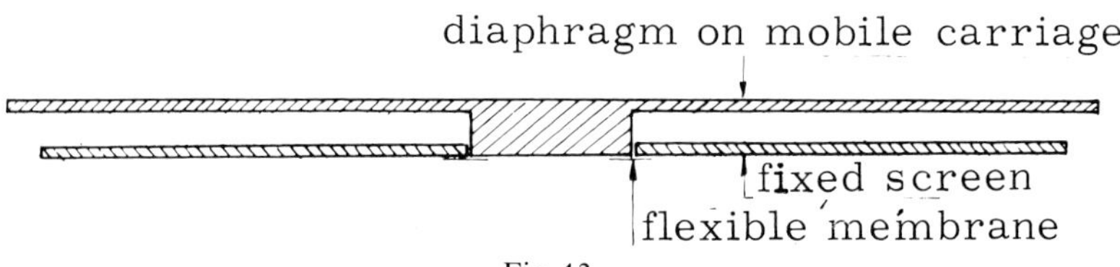

Fig. 13

4.3.1. Experimental Set-up and Procedure

For each test the mass of granular material was formed by allowing the grains of the material in question always to fall freely from a constant height of 5 cm, without any manipulation of the mass thus formed, except in those cases where it was vibrated or compacted (see Section 10).

After each load application, and before further loading, the displacements of the diaphragm pending stabilisation of the material were observed with the aid of the dial gauges.

A pull F exerted on the carriage is equilibrated by the passive resistance developed in the opposite direction by the granular material:

$$F = \frac{\gamma \cdot h^2}{2} \cdot b \cdot k_p$$

where k_p is the *passive resistance coefficient,* namely:

$$k_p = \frac{2\,F}{\gamma \cdot h^2 \cdot b}$$

This tensile force F causes the diaphragm to undergo a translatory displacement Δl.

It is therefore possible at all times to determine – for a given value of the force F acting on the carriage – the *displacement Δl* of the diaphragm and the corresponding *passive resistance coefficient* at one and the same time.

4.3.2. Note

The experiments have shown (see Section 13) that during the stages of adaptation and elasto-plastic equilibrium of the test materials the stress-strain behaviour of these materials, when loaded under passive resistance conditions, will – in the case of materials bounded by side walls, i.e., forming a mass of rectangular cross-sectional shape – remain unchanged for any further increase in the width b' in excess of three times the width b of the diaphragm ($b' \geqslant 3b$), but that this width can be reduced to only 1.2 times the width of the diaphragm ($b' \geqslant 1.2b$) in the case of a mass having a *trapezoidal* cross-sectional shape.

Thus we have a choice between these two alternative solutions for carrying out the experiments for determining the stress-strain relationships in the first two stages of equilibrium of granular materials.

5. Experiments Relating to the Surface Conditions of the Diaphragm

In Volume I of this book (1)*, as well as in the papers (2) and (3), the authors reported in detail on their experimental work showing that in the "at rest" state of equilibrium of a granular material retained by a diaphragm the surface condition of the latter had no effect on the magnitude of the horizontal component of maximum thrust and of the minimum passive resistance associated with that material.

It will be recalled that the demonstration of that principle emerged from our experiments in which a mass of material developed passive resistance on one side of the diaphragm so as to equilibrate the thrust developed by a mass of material on the other side, with a given height greater than that of the passively resisting material. It was thus verified that the equilibrium of the two masses was always exactly the same, whatever the nature of the surface of the diaphragm, even if one face thereof was rough and the other smooth, and vice versa.

It was considered necessary to ascertain the conditions in which this principle continues to be valid beyond the "at rest" state of equilibrium of the material and the diaphragm, i.e., when the diaphragm undergoes displacement and, in so doing, pushes back the material.

The experiments carried out for this purpose consisted in investigating the deformations of a granular material under the action of a diaphragm performing a translatory movement horizontally or along descending or ascending inclined planes.

The tests were executed in accordance with the procedure described in Section 4.2.

5.1. Experiments Relating to Passive Resistance Due to Horizontal Translatory Movement of the Diaphragm

All the experiments relating to the deformations Δl of a granular material

---

loaded under passive resistance conditions by a diaphragm tending to push it back upon itself were performed with the aid of a rough diaphragm coated with coarse-grained emery. Afterwards, in order to study the effect of surface condition on the magnitude of the passive resistance, the diaphragm was covered with a sheet of polished tinfoil so as to ensure that the granular materials would rub against a smooth face.

Verification of the angle of friction on a smooth face was based on the test results given in Table 1.

Applied Forces F (in kg)	$\triangle\,l$ (in 1/100 mm)
0.500	4.4
0.830	8.5
1.160	18.6
1.490	34.4
1.820	58.5
2.150	126.2
2.480	217.0
2.550	sliding
2.655	failure

Table 1

The load applied to the friction plate, including its own weight:

$$5.000 \text{ kg}$$

The load which caused first sliding movement of the plate:

$$2.550 \text{ kg}$$

corresponding to the minimum coefficient of friction:

$$\text{tg } \varphi'_{min} = \frac{2.550}{5.000} = 0.510 \quad \text{and therefore} \quad \varphi'_{min} = 27° < \alpha = 33° \; 40'$$

The load which caused final breakdown of the equilibrium of the plate (failure):

$$2.655 \text{ kg}$$

corresponding to the maximum coefficient of friction:

$$\text{tg } \varphi'_{max} = \frac{2.655}{5.000} = 0.531 \quad \text{and therefore} \quad \varphi'_{max} = 27° \; 55'$$

Experiments relating to the passive resistance in the case of a rough and a smooth diaphragm are described below:

For fine sand, for example, whose physical and mechanical properties are given in Section 3 and which is disposed in a mass with horizontal top surface and subjected to a force exerted on it (and tending to push it back upon itself) by a vertical diaphragm 0.120 m high and 0.20 m wide, the test results given in Table 2 were obtained. Each figure is the average of three consecutive tests, performed in each case on a smooth diaphragm and on a rough diaphragm.

Forces Pushing Back Material (in kg)	$\triangle l$ (in 1/100 mm)	
	smooth diaphragm	rough diaphragm
1	3.4	3.6
2	21.9	22.8
3	64.2	63.6
4	115.0	98.0
5	178.1	151.8
6	229.0	206.8
7	282.0	249.4
8	360.6	308.7
9	420.0	366.2
10	488.1	423.5
11	555.2	478.9
12	615.0	522.7
13	680.1	591.5
14	770.0	624.3
15	865.0	730.0

Table 2: Results of tests with horizontal translatory movement

Results of tests with horizontal translatory movement show that the start of the stage of adaptation defined in Section 6.1.1., from the application of the first forces to the diaphragm up to a force of about 3 kg, the deformations of the material are practically the same, for any one particular value of this force tending to push back the material, whether the diaphragm is smooth or rough.

Then, when the force pushing back the material increases and affects more directly the stage of elasto-plastic equilibrium thereof, the deformations associated with the smooth diaphragm are greater than those associated with the rough diaphragm. Hence it follows that the passive resistance reactions of the granular material are smaller in the case of the smooth, and larger in the case of the rough diaphragm.

5.2. Experiments Relating to Passive Resistance Due to Translatory Movement of the Diaphragm along Ascending or Descending Inclined Planes

In civil engineering practice it is occasionally observed that a retaining wall fails as a result of inadequate strength of the foundation supporting it.

The consequences of this in so far as they affect the magnitude of the passive resistance that the soil can then develop can be investigated on a scale model by measuring the displacements Δl of the diaphragm which correspond to definite values of the force applied to it, while at the same time causing the diaphragm to descend vertically in conjunction with its translatory movement.

For this purpose the tests reported below were carried out in circumstances where the steel balls supporting the mobile carriage described in Section 4.2. rolled on inclined planes (angle of inclination δ) with respect to the horizontal.

The material used in these tests was fine sand similar to that already referred to. The diaphragm was smooth in some of the tests, rough in others, and moved along a descending inclined plane with a slope of 6% ($\delta = 3°30'$). In Table 3 the test results are given.

Forces Pushing Back Material (in kg)	Δl (in 1/100 mm)	
	smooth diaphragm	rough diaphragm
1	2.6	2.7
2	14.4	14.6
3	50.1	42.7
4	84.1	74.6
5	130.6	112.5
6	178.2	158.2
7	227.8	207.2
8	279.3	254.1
9	324.5	302.0
10	366.5	351.3
11	437.5	400.8
12	495.0	455.1
13	555.2	519.3
14	610.0	590.1
15	688.0	645.0

Table 3

The authors also investigated the case of a diaphragm moving along an ascending inclined plane. Although the conditions in which a retaining wall will rise vertically in relation to its foundation in actual practice are hard to imagine, the tests reported below, which were performed for the sake of completeness, were similar to the preceding ones, except that now the diaphragm, while undergoing its translatory displacement, moved along an ascending inclined plane with a slope of 6% ($\delta = 3°30'$). In Table 4 the test results are given. The results in Tables 1 to 4 are shown in Figures 14 and 15.

Forces Pushing Back Material (in kg)	$\triangle l$ (in 1/100 mm) smooth diaphragm	rough diaphragm
1	2.6	2.7
2	14.4	14.6
3	50.1	42.7
4	84.1	74.6
5	130.6	112.5
6	178.2	158.2
7	227.8	207.2
8	279.3	254.1
9	324.5	302.0
10	366.5	351.3
11	437.5	400.8
12	495.0	455.1
13	555.2	519.3
14	610.0	590.1
15	688.0	645.0

Table 4

5.3. Remarks

Figures 14 and 15 show that the curves representing the displacements Δl of the diaphragm are the same in the case of a horizontal translatory movement and the case of such movement along a descending inclined plane for a given surface condition of the diaphragm, namely smooth or rough respectively.

For a given procedure with regard to the movement of the diaphragm – horizontal or inclined – the displacements of the latter are practically the same at the start of the stage of adaption of the material, corresponding to the application of very small forces tending to push it back, whereas they are smaller with the rough screen than with the smooth screen throughout the stages of elasto-plastic and of plastic equilibrium.

So it can permissibly be inferred that the surface condition of the diaphragm has no influence at all on the magnitude of the horizontal component of the passive resistance "at rest" or when the diaphragm undergoes very small amounts of displacement (of the order of 3.5/1000 of its height).

An upward vertical displacement of the – smooth or rough – diaphragm, of the order of 1/1000 of its height is sufficient to produce an increase in the magnitude of the passive resistance.

Finally, it is to be noted that the straight lines representing the deformations of the material in its stage of elasto-plastic equilibrium – as will be more fully explained in Section 6.1. – all intersect the axis of abscissae ($\Delta l = 0$) at one and the same point corresponding to the minimum passive resistance at rest.

Hence it emerges that the magnitude of the minimum passive resistance "at rest" – for zero displacement – is the same, irrespective of the surface condition of the diaphragm – smooth or rough – and irrespective of whether the diaphragm undergoes displacement horizontally or along an inclined plane.

These results were obtained from systematic tests performed on masses of granular material 12, 20, 25 and 30 cm in height, in conformity with the conclusions drawn from the investigation of the minimum dimensions of models as reported in Section 13.

o   horizontal translatory movement

+   descending translatory movement - 6%

●   ascending translatory movement + 6%

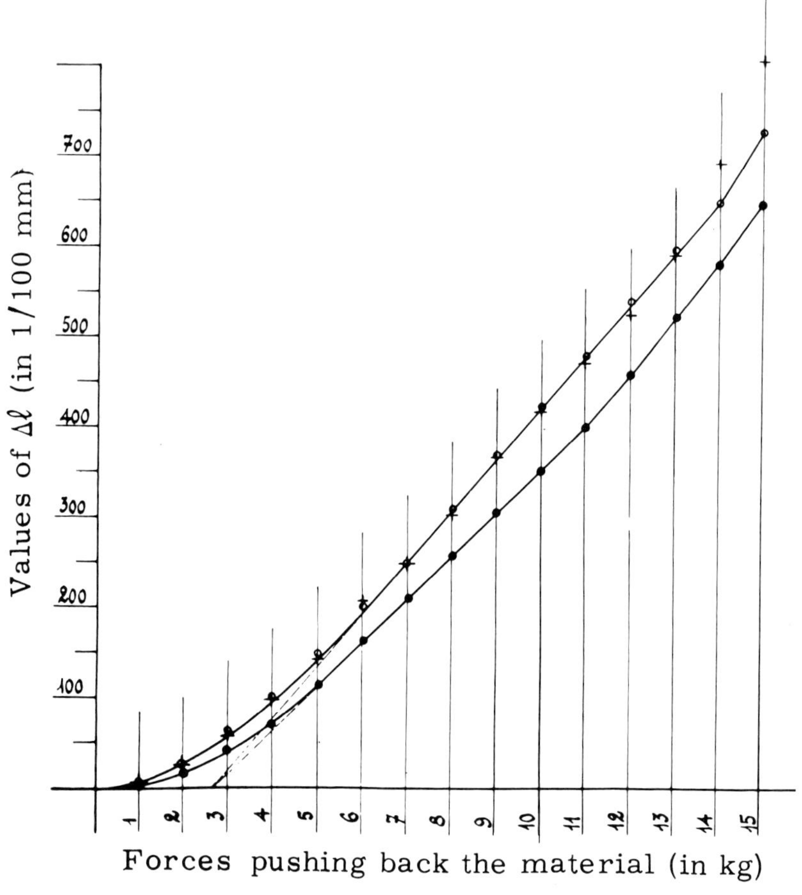

Fig. 14: Results for smooth diaphragms.
Material: Fine sand.

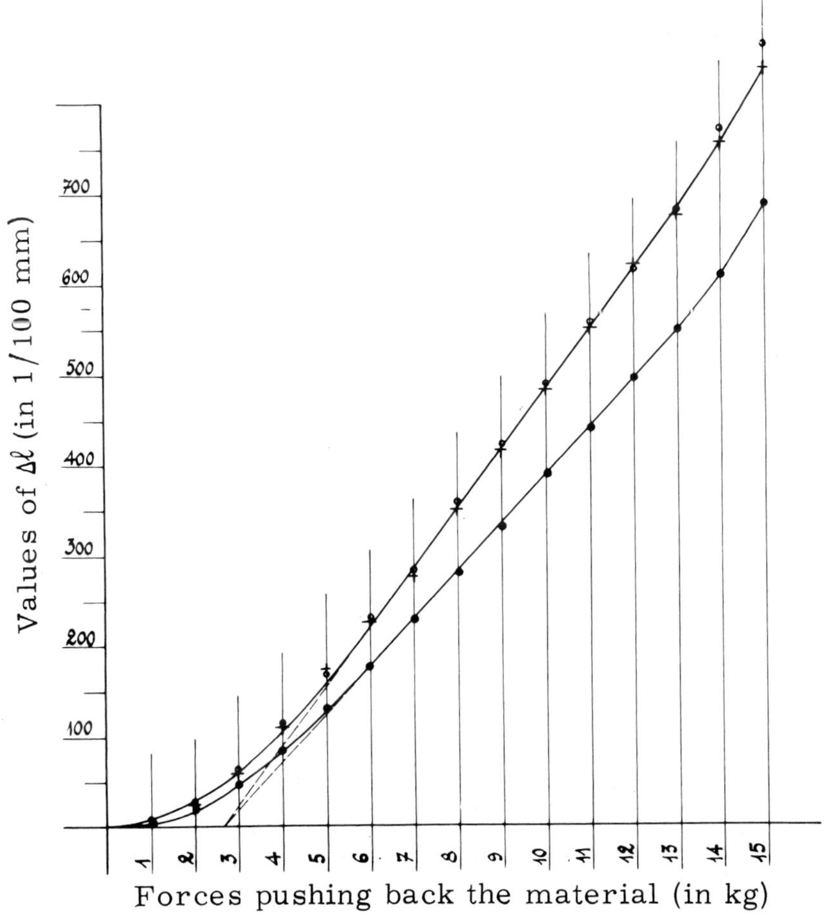

Fig. 15: Results for rough diaphragms.
Material: Fine sand.

6. Experiments Relating to Translatory Passive Resistance

The expressions for the relative values of maximum thrust and minimum passive resistance, for zero displacement of the diaphragm, were established for what have been referred to as "at rest" states of equilibrium. However, it emerged that there were cases where it was necessary to have knowledge of the phenomena occurring beyond the state of rest of the granular material if displacements of the retaining diaphragms (or in general: the retaining walls) might develop, if even only temporarily, e. g., when sheet piling is used as a temporary retaining structure while civil engineering work is being carried out.

Accordingly, fresh experimental work was undertaken with the granular materials described in Section 3 and possessing distinctly different mechanical and other properties in comparison with one another. Also, the diaphragms employed in the tests differed in height. In due course it was found that the phenomenon of the deformation of a granular material under the action of a diaphragm tending to push it back manifested itself in three well-defined characteristic stages.

6.1. Characteristic Stages of Deformation of a Granular Mass Developing Translatory Passive Resistance

The deformation of a mass of granular material subjected to passive resistance loading conditions is characterised by:

– A stage of *adaptation* of the equilibrium of the material, during which the deformations are the same as those affecting any granular material subjected to compressive forces.

– A stage of *elasto-plastic* equilibrium of the granular material, characterised by a steady increase in the deformations of the latter as a function of the forces striving to push back the material. This continues up to a limit where the corresponding passive resistance coefficient attains approximately the value

$$\frac{\pi + 2\,\varphi_0}{\pi - 2\,\varphi_0}$$

which was defined already in our earlier investigation of the "at rest" equilibrium of granular material retained by a diaphragm (or in general: by a retaining wall) as being the ratio of the value of the *permissible maximum passive resistance in structural stability calculations* to the value of the *relative minimum passive resistance,* for zero displacement, corresponding to a passive resistance coefficient equal to unity.

– A stage of *plastic* equilibrium of the material, during which its deformations increase irregularly up to failure.

6.1.1. First Stage: Adaptation of Equilibrium

During the first stage the – relatively small – deformations of the mass of granular material undergo a steady increase in accordance with a regular curve which joins the straight line representing the deformations in the next stage, i.e., that of elasto-plastic equilibrium.

6.1.2. Second Stage: Elasto-Plastic Equilibrium

In this second stage of equilibrium a notable and important feature is observed, namely, that the deformations are rigorously represented by a straight line which, when extended, intersects the axis of abscissae at a characteristic point corresponding exactly to the value of minimum passive resistance for zero displacement which will henceforth be referred to as the "relative minimum passive resistance".

6.1.3. Third Stage: Plastic Deformations of the Mass

During this stage it is observed that, once the passive resistance coefficient corresponding to the forces tending to push back the retaining diaphragm has reached a value of about

$$\frac{\pi + 2\,\varphi_0}{\pi - 2\,\varphi_0}$$

the displacements of the diaphragm thereafter increase irregularly with respect to the applied loads and that this stage of behaviour is not susceptible of any interpretation whereby the experimentally obtained results can be expressed in formulas. Accordingly, this stage will only be briefly referred to in the further experimental work which will be devoted to an exhaustive investigation of the subject within the limits of the first two stages of equilibrium of the granular materials tested.

In Figure 16 the curve representing the deformations occurring in the three above-mentioned stages is shown schematically:

- 0 to A: the so-called stage of adaptation of the granular material used in the experiment;
- A to B: the stage of elasto-plastic equilibrium;
- B to C: the stage of plastic equilibrium of the mass of granular material up to failure.

By way of example, some test results obtained with 0.15 m high diaphragms will be given further on.

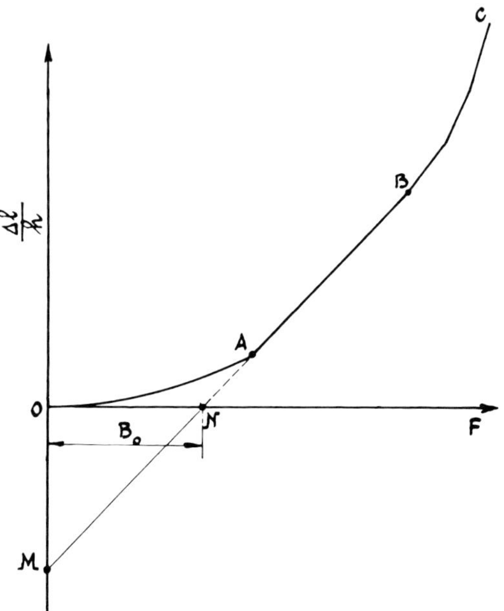

Fig. 16: F = Forces pushing back the material
B_0 = Relative minimum passive resistance

6.1.4. Note on Relative Minimum Passive Resistance for Zero Displacement

According to the experiments referred to in Sections 6.3. and 6.4., relating to granular materials with top surface inclined at the upper $(+\alpha)$ or the lower angle of repose $(-\alpha)$ – or, in a more general way, at any angle α' not exceeding α – we shall see (as has already been indicated in a preliminary manner in Section 6.1.2.) that the straight line representing the deformations in the stage of elasto-plastic equilibrium will, when extended, intersect the axis of abscissae at a point which, generally speaking, corresponds to the relative minimum passive resistance B_0 for zero displacement – i.e., it corresponds to the passive resistance coefficient

$$1 \pm \frac{2\,\alpha'}{\pi}$$

Hence:

a) In the case of material with horizontal top surface and a width b:

$$\mathbf{B_0} = \frac{\gamma \cdot h^2}{2} \cdot b$$

b) In the case of material having its top surface inclined at the upper angle of repose $(+\alpha)$:

$$\mathbf{B_0'} = \frac{\gamma \cdot h^2}{2} \cdot b \left(1 + \frac{2\,\alpha}{\pi}\right)$$

c) In the case of material having its top surface inclined at the lower angle of repose $(-\alpha)$:

$$\mathbf{B_0''} = \frac{\gamma \cdot h^2}{2} \cdot b \left(1 - \frac{2\,\alpha}{\pi}\right)$$

d) In the case of material whose top surface is inclined at any angle $(\alpha' \leqslant \alpha)$ above or below the horizontal:

$$\mathbf{B_0'''} = \frac{\gamma \cdot h^2}{2} \cdot b \left(1 \pm \frac{2\,\alpha'}{\pi}\right)$$

6.2. Material with Horizontal Top Surface

6.2.1. Fine Sand

Bulk density $\qquad \gamma = 1\,380 \text{ kg/m}^3$

Angle $\varphi_0 \qquad \varphi_0 \cong \alpha = 33° \, 40'$

Vertical retaining diaphragm: Height $h = 0.15$ m
$\qquad\qquad\qquad\qquad\qquad$ Width $b = 0.20$ m

Relative minimum translatory passive resistance:

$$\mathbf{B_0} = \frac{1\,380 \times \overline{0.15^2}}{2} \times 0.20 = 3.100 \text{ kg}$$

Test results are given in Table 5 and Figure 17.

Forces Pushing Back Material (in kg)	Values of k_p (1)	Values of $\triangle$ 1 (in 1/100 mm)	Values of $\dfrac{\triangle 1}{h}$ (in thousandths)
0.600	0.193	0.4	0.026
0.800	0.258	0.9	0.06
1.000	0.322	1.3	0.086
1.200	0.387	1.7	0.113
1.400	0.452	2.1	0.14
1.600	0.516	2.6	0.173
1.800	0.58	3.2	0.213
2.000	0.645	3.9	0.26
2.200	0.71	4.7	0.313
2.400	0.775	5.6	0.373
2.600	0.838	8.9	0.593
2.800	0.902	12.7	0.848
3.000	0.967	16.7	1.115
3.200	1.031	20.4	1.36
3.400	1.098	25.3	1.69
3.600	1.16	30.0	2.0
3.800	1.225	35.2	2.35
4.000	1.29	41.0	2.73
4.200	1.355	48.2	3.22
4.400	1.42	56.2	3.75
4.600	1.485	64.5	4.40
4.800	1.55	74.2	4.95
5.000	1.615	83.4	5.6
5.200	1.68	92.2	6.15
5.400	1.74	100.5	6.7
5.600	1.805	109.5	7.3
5.800	1.87	118.6	7.92
6.000	1.935	127.6	8.51
6.200	2.0	136.5	9.1
6.400	2.065	147.5	9.83
6.600	2.13	161.0	10.72

Table 5

(1) $\quad k_p = \dfrac{F}{3.100}$

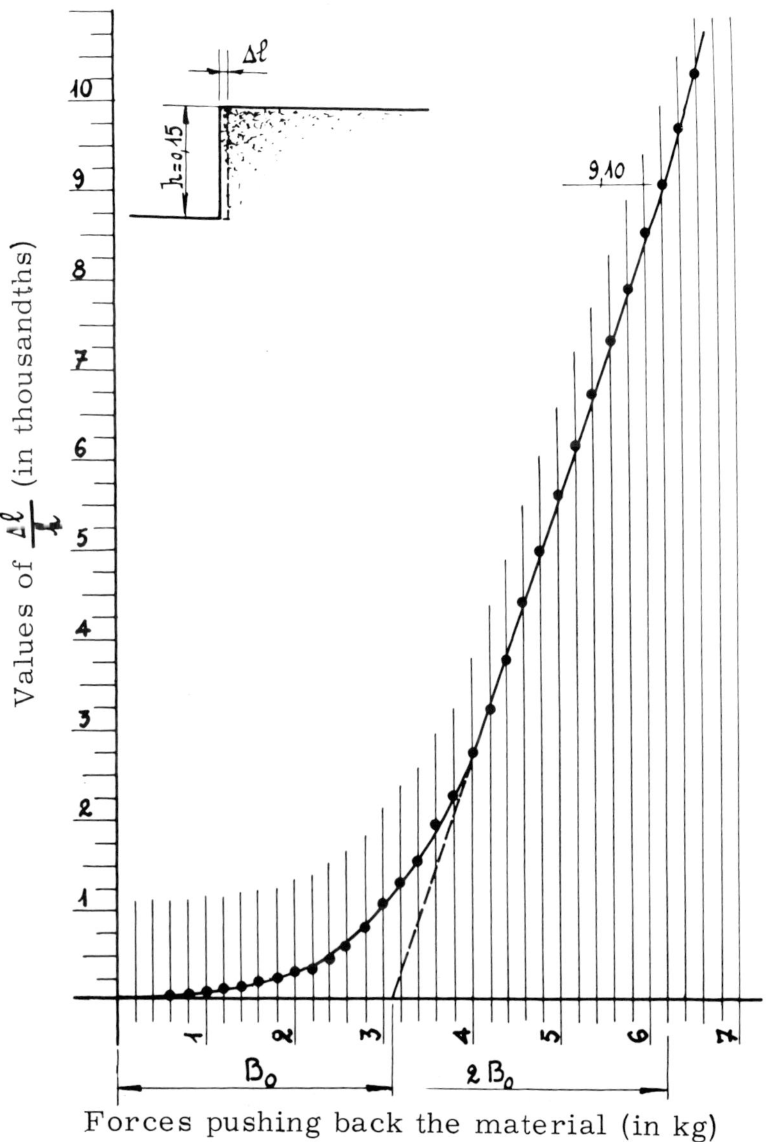

Forces pushing back the material (in kg)

Fig. 17: Material with horizontal top surface: Fine sand.
Graphical presentation of Table 5.

6.2.2. Var Crushed Stone Sand

Bulk density $\qquad\qquad \gamma = 1\,420$ kg/m³

Angle $\varphi_0 \qquad\qquad \varphi_0 \cong \alpha = 36° \, 30'$

Vertical retaining diaphragm: Height $h = 0.15$ m
Width $b = 0.20$ m

Relative minimum translatory passive resistance:

$$\mathbf{B_0} = \frac{1\,420 \times 0.15^2}{2} \times 0.20 = 3.195 \text{ kg}$$

Test results are given in Table 6 and Figure 18.

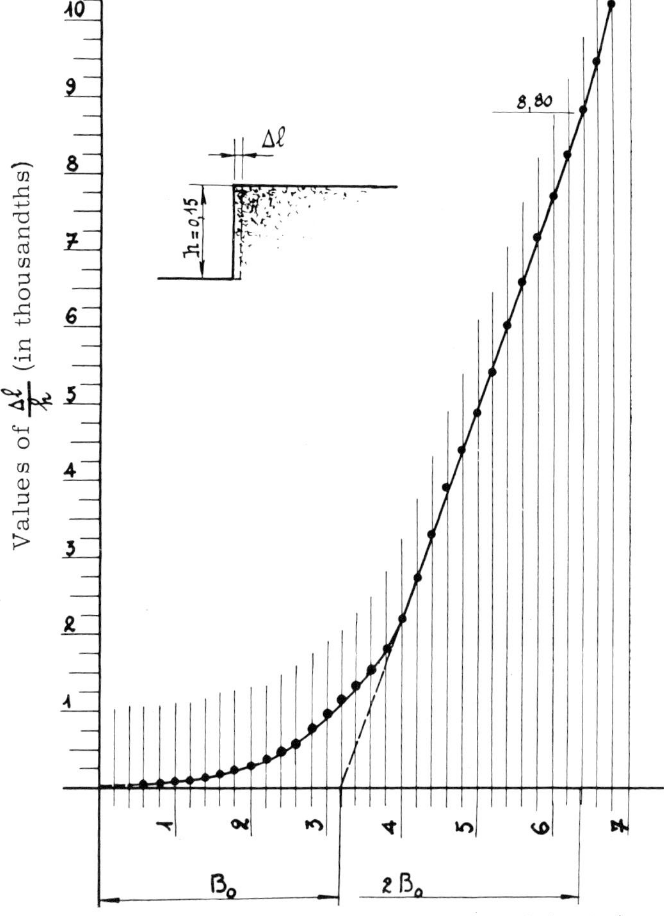

Forces pushing back the material (in kg)

Fig. 18: Material with horizontal top surface: Var crushed stone sand.
Graphical presentation of Table 6.

Applied Forces F (in kg)	Values of k_p (1)	Values of $\triangle l$ (in 1/100 mm)	Values of $\dfrac{\triangle l}{h}$ (in thousandths)
0.600	0.188	1.05	0.07
0.800	0.25	1.5	0.10
1.000	0.313	1.9	0.13
1.200	0.375	2.4	0.16
1.400	0.438	3.0	0.20
1.600	0.501	3.7	0.25
1.800	0.563	4.5	0.30
2.000	0.626	5.7	0.38
2.200	0.688	6.9	0.46
2.400	0.751	8.6	0.58
2.600	0.814	10.5	0.70
2.800	0.876	12.7	0.85
3.000	0.94	15.3	1.02
3.200	1.001	18.1	1.21
3.400	1.064	21.6	1.44
3.600	1.127	25.5	1.70
3.800	1.189	29.1	1.94
4.000	1.252	33.6	2.24
4.200	1.315	41.1	2.74
4.400	1.378	49.5	3.30
4.600	1.44	57.9	3.86
4.800	1.502	66.3	4.42
5.000	1.567	73.2	4.88
5.200	1.628	81.3	5.42
5.400	1.69	90.6	6.04
5.600	1.753	99.0	6.60
5.800	1.815	107.4	7.16
6.000	1.878	115.5	7.70
6.200	1.94	123.6	8.24
6,400	2.003	132.0	8.80
6.600	2.066	141.1	9.40
6.800	2.128	151.6	10.10
7.000	2.191	163.5	10.90

Table 6

(1) $k_p = \dfrac{F}{3.195}$

6.2.3. Seine Sand

Bulk density $\qquad\qquad\qquad\gamma = 1\,550\ \text{kg/m}^3$

Angle φ_0 $\qquad\qquad\qquad \varphi_0 \cong \alpha = 34°\ 40'$

Vertical retaining diaphragm: Height $h = 0.15$ m
$\qquad\qquad\qquad\qquad\qquad$ Width $\ \ b = 0.20$ m

Relative minimum translatory passive resistance:

$$B_0 = \frac{1\,550 \times 0.15^2}{2} \times 0.20 = 3.490\ \text{kg}$$

Test results are given in Table 7 and Figure 19.

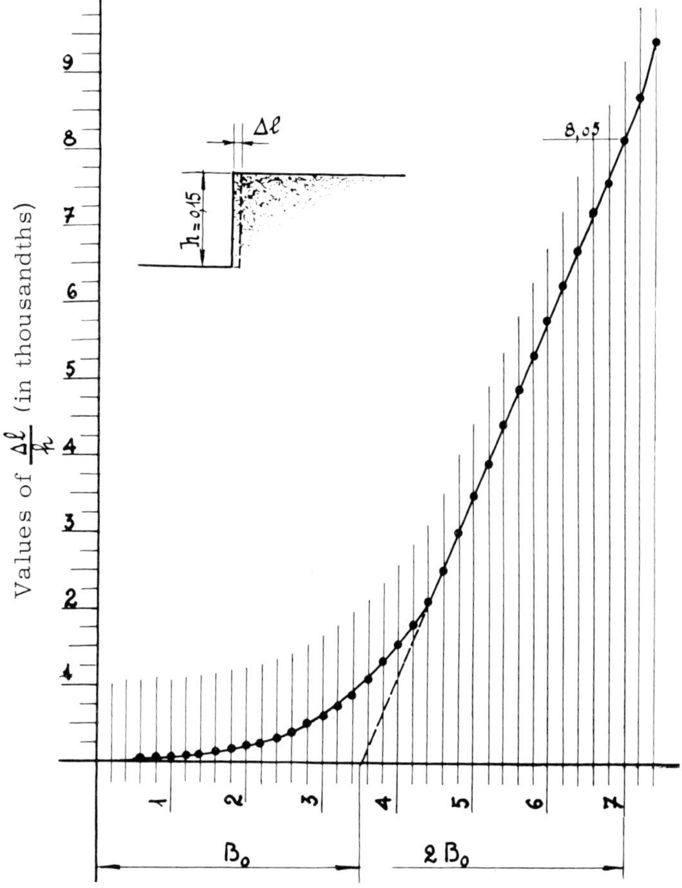

Forces pushing back the material (in kg)

Fig. 19: Material with horizontal top surface: Seine sand.
Graphical presentation of Table 7.

Applied Forces F (in kg)	Values of k_p (1)	Values of $\triangle$ 1 (in 1/100 mm)	Values of $\dfrac{\triangle 1}{h}$
0.600	0.172	0.9	0.06
0.800	0.23	1.2	0.08
1.000	0.289	1.5	0.10
1.200	0.344	1.8	0.12
1.400	0.401	2.1	0.14
1.600	0.458	2.7	0.18
1.800	0.516	3.31	0.22
2.000	0.573	3.9	0.26
2.200	0.63	4.52	0.30
2.400	0.687	5.39	0.36
2.600	0.745	6.65	0.43
2.800	0.802	7.92	0.53
3.000	0.86	9.3	0.62
3.200	0.917	11.12	0.74
3.400	0.974	13.2	0.88
3.600	1.031	17.11	1.14
3.800	1,089	20.7	1.38
4.000	1.146	24.1	1.60
4.200	1.203	27.9	1.86
4.400	1.26	31.82	2.12
4.600	1.318	38.1	2.54
4.800	1.375	45.6	3.04
5.000	1.432	52.5	3.50
5.200	1.49	58.4	3.90
5.400	1.547	66.8	4.45
5.600	1.604	73.2	4.88
5.800	1.662	79.91	5.33
6.000	1.72	87.1	5.80
6.200	1.777	93.76	6.25
6.400	1.834	100.5	6.70
6.600	1.891	107.7	7.18
6.800	1.948	111.1	7.41
7.000	2.006	121.6	8.10
7.200	2.063	129.1	8.60
7.400	2.12	140.2	9.35

Table 7

(1) $k_p = \dfrac{F}{3.490}$

6.2.4. Millet

Bulk density $\gamma = 730 \text{ kg/m}^3$

Angle φ_0 $\varphi_0 \cong \alpha = 28° \; 30'$

Top surface of the material is horizontal.

Vertical retaining diaphragm: Height $h = 0.15$ m
                                           Width $b = 0.20$ m

Minimum translatory passive resistance:

$$\mathbf{B_0} = \frac{730 \times \overline{0.15}^2}{2} \times 0.20 \cong 1.640 \text{ kg}$$

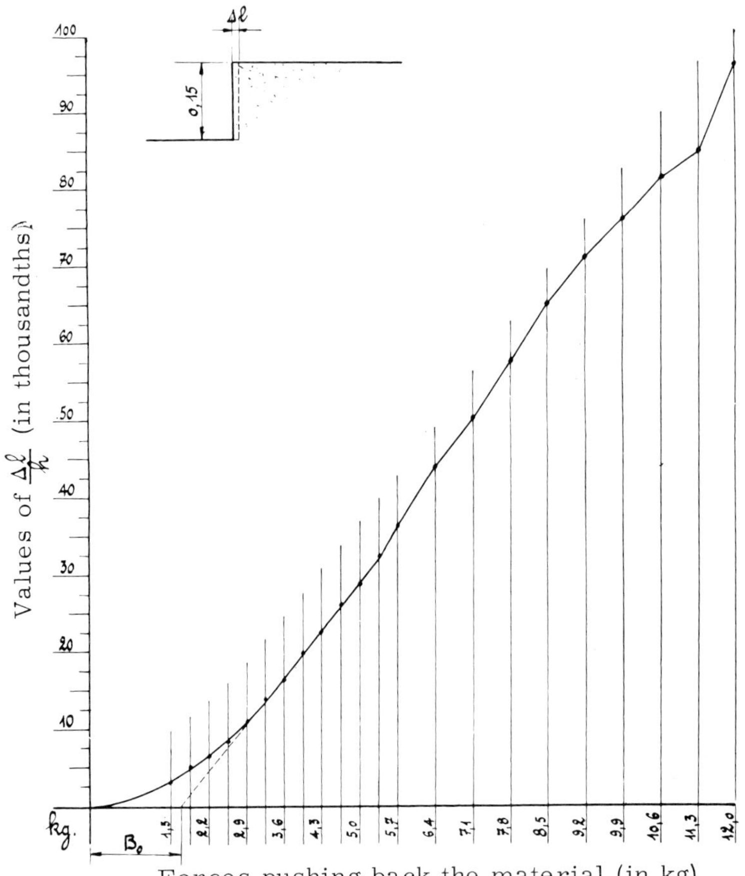

Forces pushing back the material (in kg)

Fig. 20: Translatory passive resistance: Millet.
Graphical presentation of Table 8.

Test results are given in Table 8 and Figure 20.

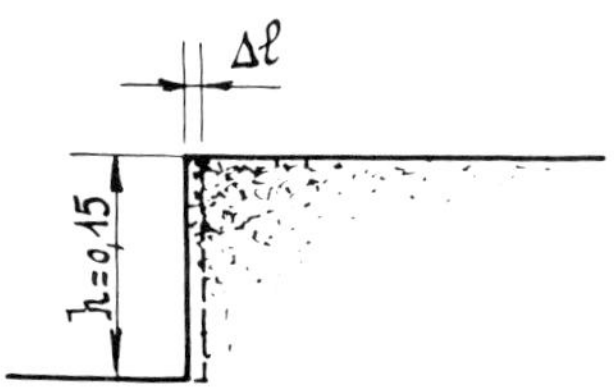

Forces Applied Succesively (in kg)	Applied Forces F (in kg)	Values of k_p (1)	Values of $\triangle\,1$ (in 1/100 mm)	Values of $\dfrac{\triangle\,1}{h}$ (in thousandths)
1.500	1.500	0.915	45.0	3
0.350	1.850	1.13	75.0	5
»	2.200	1.34	97.5	6.5
»	2.550	1.555	126.0	8.4
»	2.900	1.77	165.0	11.0
»	3.250	1.98	208.5	13.9
»	3.600	2.195	246.0	16.4
»	3.950	2.41	294.0	19.6
»	4.300	2.62	334.0	22.4
»	4.650	2.835	393.0	26.2
»	5.000	3.05	427.5	28.5
»	5.350	3.26	480.0	32.0
0.700	5.700	3.435	547.0	36.5
»	6.400	3.90	645.0	43.0
»	7.100	4.325	750.0	50.0
»	7.800	4.75	860.0	57.4
»	8.500	5.175	975.0	65.0
»	9.200	5.61	1 055.0	70.3
»	9.900	6.03	1 140.0	76.0
»	10.600	6.46	1 230.0	82.0
»	11.300	6.89	1 264.0	84.2
»	12.000	7.31	1 435.0	95.6

(1) $\quad k_p = \dfrac{F}{1.640}$

Table 8

6.2.5. Wheat

Bulk density $\qquad\qquad\qquad \gamma = 790 \ \text{kg/m}^3$

Angle $\varphi_0 \qquad\qquad \varphi_0 \cong \alpha = 31° \ 30'$

Top surface of the material is horizontal.

Vertical retaining diaphragm: Height $h = 0.15$ m
Width $b = 0.20$ m

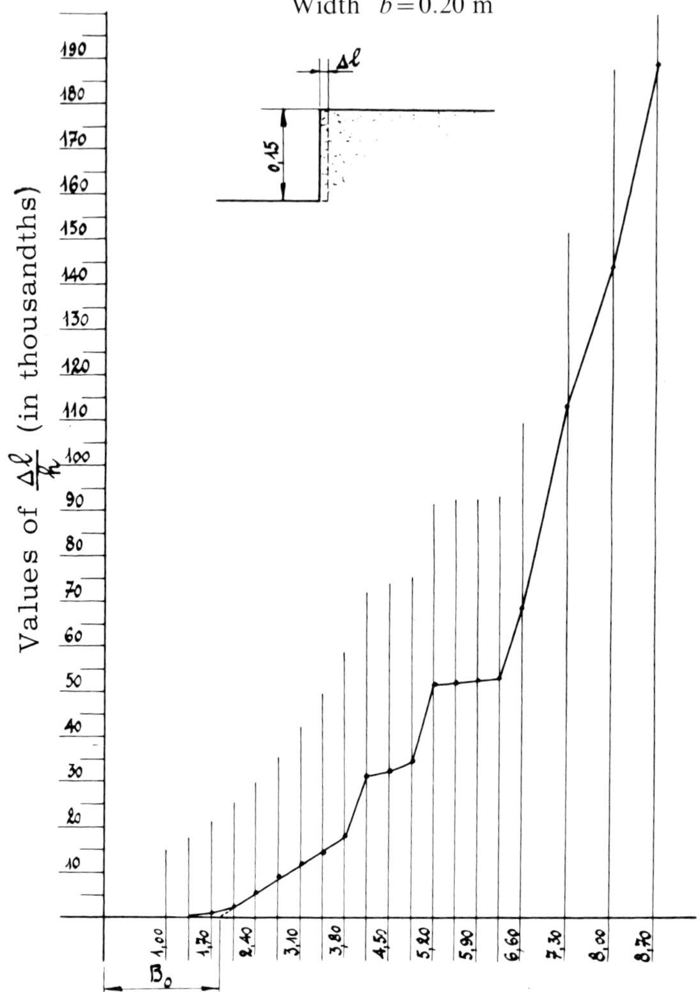

Fig. 21: Translatory passive resistance: Wheat.
Graphical presentation of Table 9.

Minimum translatory passive resistance:

$$\mathbf{B}_0 = \frac{790 \times \overline{0.15^2}}{2} \times 0.20 \cong 1.780 \text{ kg}$$

Test results are given in Table 9
and Figure 21.

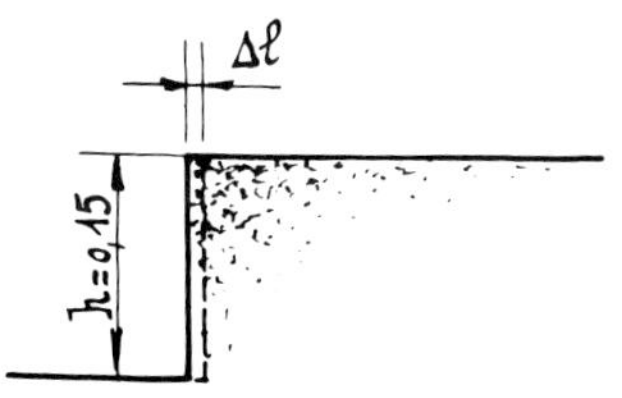

Forces Applied Succesively (in kg)	Applied Forces F (in kg)	Values of k_p (1)	Values of $\triangle$ l (in 1/100 mm)	Values of $\frac{\triangle\ l}{h}$ (in thousandths)
1.000	1.000	0.562	0.1	0.006
0.350	1.350	0.76	2	0.134
»	1.700	0.955	14.5	0.97
»	2.050	1.15	36	2.4
»	2.400	1.35	76.5	5.1
»	2.750	1.525	147	9.8
»	3.100	1.74	178	11.85
»	3.450	1.94	205	13.65
»	3.800	2.13	259	17.3
»	4.150	2.325	467	31.1
»	4.500	2.625	482	32.1
»	4.850	2.72	520	34.6
»	5.200	2.92	768	51.2
»	5.550	3.12	770	51.3
»	5.900	3.31	775	51.6
»	6.250	3.51	785	52.3
0.700	6.600	3.71	1 060	68.75
»	7.300	4.10	1 680	112
»	8.000	4.50	2 150	143
»	8.700	4.89	2 820	188
1.000	9.400	5.28	measurements interrupted,	
»	10.400	5.60	very variable from test to test	
»	11.400	6.40		
»	12.400	6.98		
»	13.400	7.52	slip of the diaphragm	

Table 9

(1) $k_p = \dfrac{F}{1.780}$

6.3. Material with Top Surface Inclined at the Upper Angle of Repose $(+\alpha)$

Analysis of the preceding test results, and of the results obtained in the investigation of deformation phenomena relating to retained materials with top surface inclined at angles corresponding to the upper $(+\alpha)$ or the lower angle of repose $(-\alpha)$, shows all these tests to reveal the same equilibrium phenomena, i.e., first adaptation, then elasto-plastic equilibrium and finally plastic equilibrium, as in the case of materials having a horizontal top surface.

Hence it should be possible to limit the following experiments to the first two stages referred to, namely, adaptation and elasto-plastic equilibrium (for which more detailed experimental investigations can be envisaged, as will be seen in Section 9), and to confine them to a choice of, say, two materials, e.g., fine sand from Garoupe to represent the sands, and millet to represent the cereals.

The results obtained for the translatory passive resistance developed by materials having the top surface inclined at the upper angle of repose $(+\alpha)$, more particularly in tests with fine sand and with millet, are as follows.

6.3.1. Fine Sand

Bulk density $\qquad \gamma = 1\ 380\ \text{kg/m}^3$

Minimum angle of friction $\qquad \varphi_0 \cong \alpha = 33°\ 40'$

Top surface of the material is inclined at upper angle of repose: $+\alpha$.

Vertical retaining diaphragm: Height $h=0.15$ m,
$\qquad\qquad\qquad\qquad\qquad$ Width $b=0.20$ m.

Relative minimum translatory passive resistance:

$$\mathbf{B_0} = \frac{1\ 380 \times \overline{0.15}^2}{2} \times 0.20 \cdot \left(1 + \frac{2 \times 33°\ 40'}{180°}\right) = 4.270\ \text{kg}$$

Test results are given in Table 10 and Figure 22.

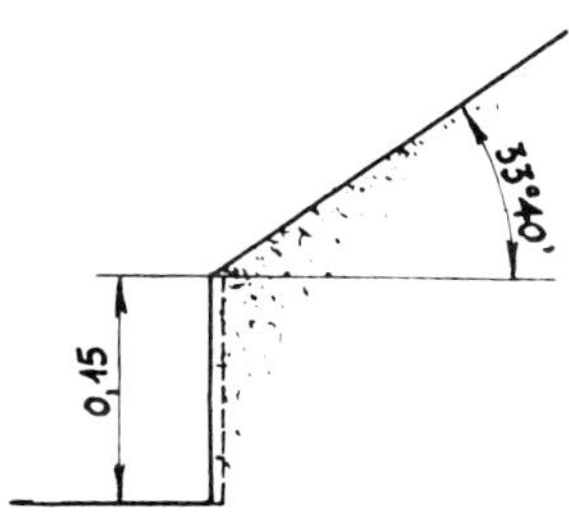

Applied Forces F (in kg)	Values of k_p (1)	Values of $\triangle$ l (in 1/100 mm)	Values of $\dfrac{\triangle\,l}{h}$ (in thousandths)
3.000	0.700	6.7	0.45
3.200	0.747	8.1	0.54
3.400	0.794	9.1	0.6
3.600	0.841	10.5	0.7
3.800	0.880	11.8	0.79
4.000	0.937	14.2	0.95
4.200	0.980	15.7	1.05
4.400	1.028	17.4	1.16
4.600	1.075	19.5	1.30
4.800	1.121	21.7	1.45
5.000	1.168	23.4	1.56
5.200	1.215	28.6	1.91
5.400	1.261	34.5	2.30
5.600	1.308	41.7	2.78
5.800	1.355	47.1	3.12
6.000	1.401	53.8	3.59
6.200	1.448	61.1	4.07
6.400	1.495	67.5	4.50
6.600	1.542	73.6	4.88
6.800	1.588	79.7	5.32
7.000	1.635	86.8	5.78
7.200	1.682	101.2	6.75

Table 10

---

(1) $k_p = \dfrac{F}{4.270}$

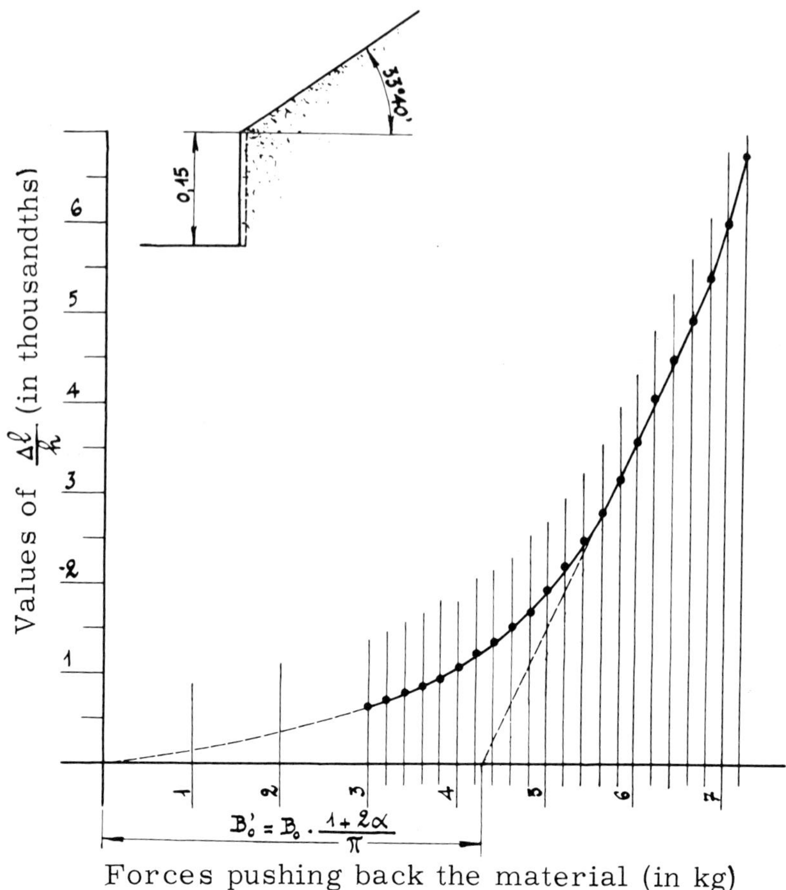

Forces pushing back the material (in kg)

Fig. 22: Material with top surface inclined at the upper angle of repose: Fine sand. Graphical presentation of Table 10.

6.3.2. Millet

Bulk density $\qquad\qquad \gamma = 730$ kg/m^3

Angle φ_0: $\qquad\qquad \varphi_0 \cong \alpha = 28° \ 30'$

Top surface of the material is inclined at upper angle of repose: $+\alpha$.

Vertical retaining diaphragm: Height $h = 0.15$ m,
$\qquad\qquad\qquad\qquad\qquad\qquad$ Width $b = 0.20$ m.

Minimum translatory passive resistance:

$$B'_0 = \frac{730 \times \overline{0.15}^2}{2} \times 0.20 \cdot \left(1 + \frac{2 \times 28° \, 30'}{180°}\right) = 2.160 \text{ kg}$$

Test results are given in Table 11 and Figure 23.

Forces Applied Succesively (in kg)	Applied Forces F (in kg)	Values of k_p (1)	Values of $\triangle 1$ (in 1/100 mm)	Values of $\frac{\triangle 1}{h}$ (in thousandths)
1.500	1.500	0.695	10.0	0.666
0.350	1.850	0.855	15.0	1.0
»	2.200	1.02	24.0	1.6
»	2.550	1.18	41.0	2.73
»	2.900	1.385	72.0	4.8
»	3.250	1.505	112.9	7.52
»	3.600	1.666	149.5	10.0
»	3.950	1.83	180.0	12.0
»	4.300	1.99	217.5	14.5
»	4.650	2.15	240.0	16.0
»	5.000	2.315	285.0	19.0
»	5.350	2.48	315.0	21.0
0.700	5.700	2.64	368.0	24.5
»	6.400	2.96	496.0	33.0
»	7.100	3.285	570.5	38.0
»	7.800	3.61	667.5	44.5
»	8.500	3.935	735.0	49.0
»	9.200	4.26	826.2	55.0
»	9.200	4.26	826.2	55.0
»	9.900	4.58	892.5	59.5
»	10.600	4.90	990.0	66.0
»	11.300	5.23	1 080.0	72.0
»	12.000	5.55	1 110.0	74.0

(1) $k_p = \dfrac{F}{2.160}$

Table 11

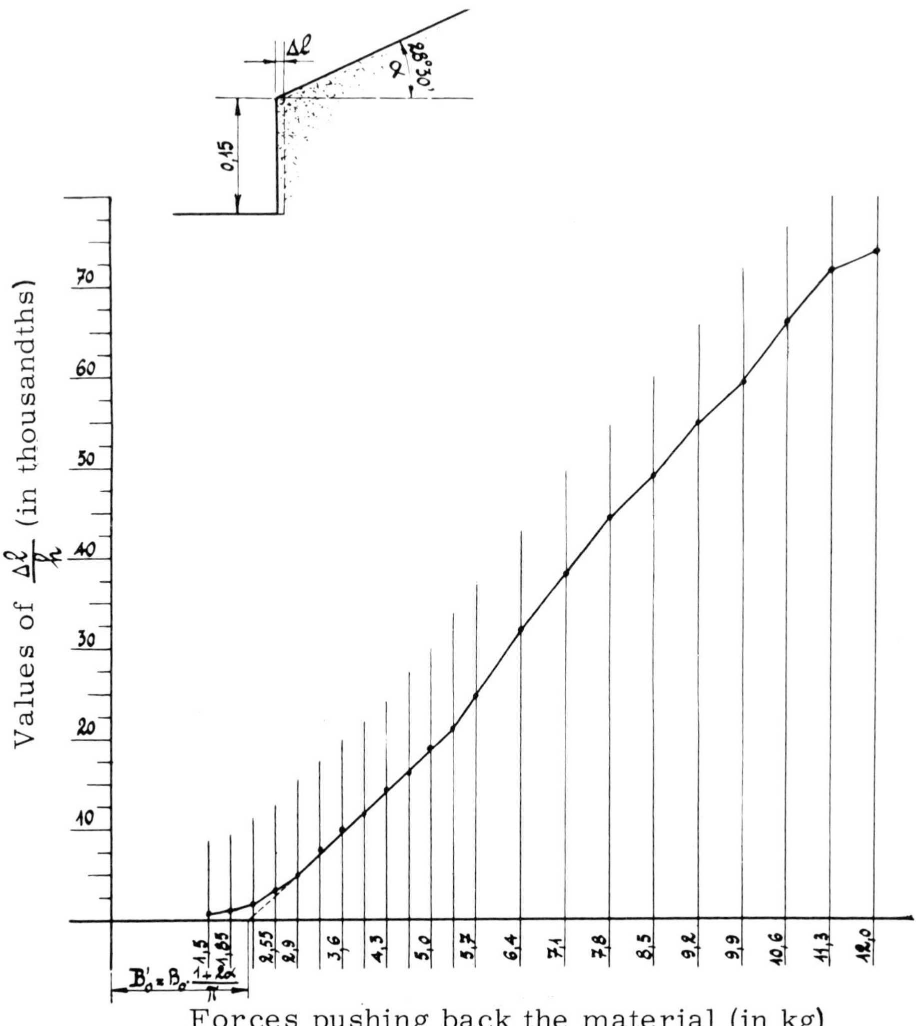

Fig. 23: Translatory passive resistance: Millet.
Graphical presentation of Table 11.

6.4. Material with Top Surface Inclined at the Lower Angle of Repose $(-\alpha)$

The results obtained for the translatory passive resistance developed by materials with top surface inclined at the lower angle of repose $(-\alpha)$, more particularly in tests with fine sand and with millet, are as follows.

6.4.1. Fine Sand

Bulk density $\qquad \gamma = 1\ 380\ \mathrm{kg/m^3}$

Minimum angle of friction $\quad \varphi_0 \cong \alpha = 33°\ 40'$

Top surface of the material is inclined at lower angle of repose: $-\alpha$.

Vertical retaining diaphragm: Height $h = 0.15$ m,
$\qquad\qquad\qquad\qquad\qquad$ Width $b = 0.20$ m.

Relative minimum translatory passive resistance:

$$B_0 = \frac{1\ 380 \times \overline{0.15}^2}{2} \times 0.20 \cdot \left(1 - \frac{2 \times 33°\ 40'}{180°}\right) = 1.945\ \mathrm{kg}$$

Test results are given in Table 12 and Figure 24.

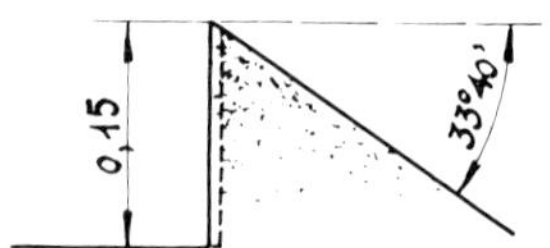

Applied Forces F (in kg)	Values of k_p (1)	Values of $\triangle\, l$ (in 1/100 mm)	Values of $\dfrac{\triangle\, l}{h}$ (in thousandths)
1.800	0.925	18.1	1.2
2.000	1.028	22.4	1.5
2.200	1.131	28.6	1.9
2.400	1.234	36.1	2.4
2.600	1.337	47.3	3.15
2.800	1.440	60.8	4.05
3.000	1.543	74.3	4.95
3.200	1.645	87.1	5.8
3.400	1.748	101.3	6.75
3.600	1.850	114.6	7.65
3.800	1.953	129.6	8.65
4.000	2.056	141.7	9.45
4.200	2.159	156.8	10.45
4.400	2.262	172.5	11.5
4.600	2.365	198.1	13.2
4.800	2.648	232.0	15.45
5.000	2.517	286.2	19.1

(1) $\quad k_p = \dfrac{F}{1.945}$

Table 12

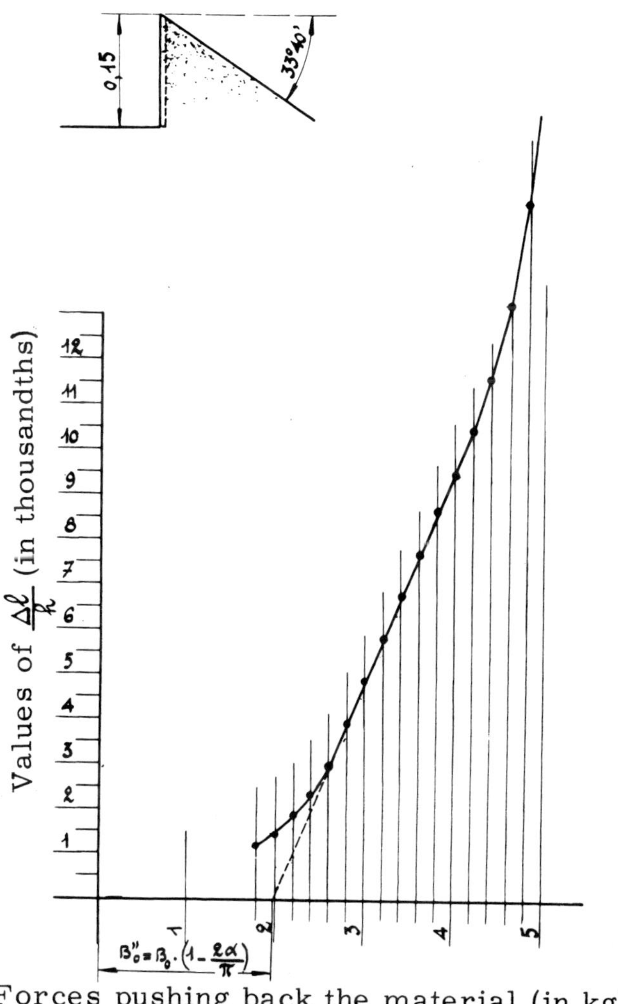

Fig. 24: Material with top surface inclined at the lower angle of repose: Fine sand. Graphical presentation of Table 12.

6.4.2. Millet

Bulk density	$\gamma = 730 \ \text{kg/m}^3$
Angle φ_0	$\varphi_0 \cong \alpha = 28°\ 30'$

Top surface of the material is inclined at lower angle of repose: $-\alpha$.

Vertical retaining diaphragm: Height $h = 0.15$ m,
                                    Width $b = 0.20$ m.

Minimum translatory passive resistance:

$$B_0'' = \frac{730 \times \overline{0.15^2}}{2} \times 0.20 \cdot \left(1 - \frac{2 \times 28°\ 30'}{180°}\right) \cong 1.125 \ \text{kg}$$

Test results are given in Table 13 and Figure 25.

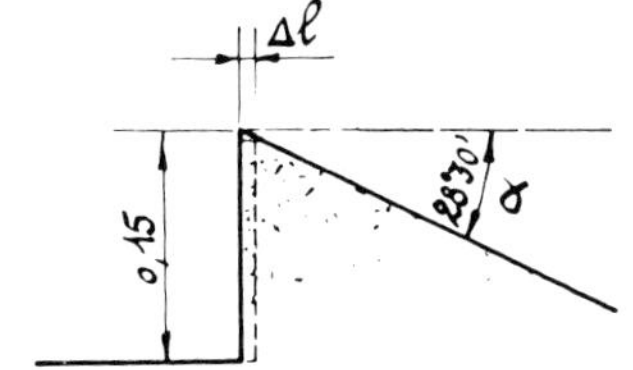

Forces Applied Succesively (in kg)	Cumulative Forces F (in kg)	Values of k_p (1)	Values of $\triangle l$ (in 1/100 mm)	Values of $\frac{\triangle l}{h}$ (in thousandths)
1.500	1.500	1.333	82.5	5.5
0.350	1.850	1.645	142.5	9.5
»	2.200	1.955	210	14
»	2.550	2.265	270,1	18
»	2.900	2.58	345.2	23
»	3.250	2.885	450	30
»	3.600	3.2	599.8	40
»	3.950	3.51	840	56
»	4.300	3.82	1 015	71
»	4.650	4.13	1 532	102
»	5.000	4.44	2 110	141
»	5.350	4.75	2 940	196
0.700	5.700	5.06	3 020	201
»	6.400	5.69	3 910	261
	7.100	6.31		sliding

(1) $k_p = \dfrac{F}{1.125}$

Table 13

Note:

In the case of millet whose top surface is inclined at the lower angle of repose the deformations of the material rapidly become irregular because of dilatancy which causes grains to slide down the slope and increase its steepness.

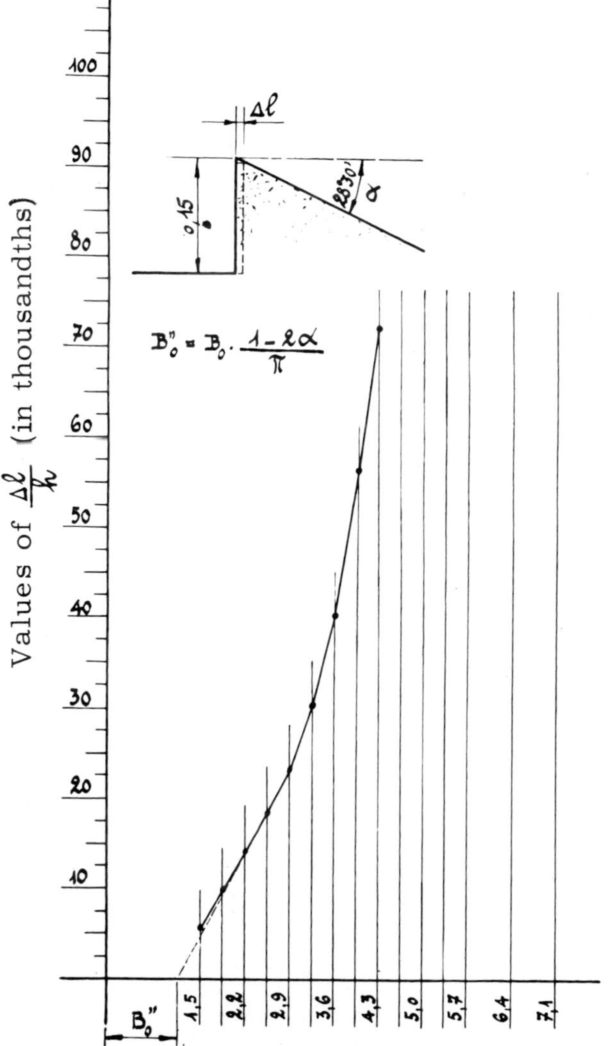

Forces pushing back the material (in kg)

Fig. 25: Translatory passive resistance: Millet.
Graphical presentation of Table 13.

6.5. Conclusions Relating to Material with Horizontal or Inclined Top Surface

The foregoing results for a mass of granular material with a horizontal or inclined top surface can be summarised in Figure 26 relating to tests performed with fine sand.

This graph clearly reveals the particular character of the straight lines representing the deformations of material during the elasto-plastic stage of equilibrium. These lines intersect the axis of abscissae at points which correspond respectively to the value of the relative minimum passive re-

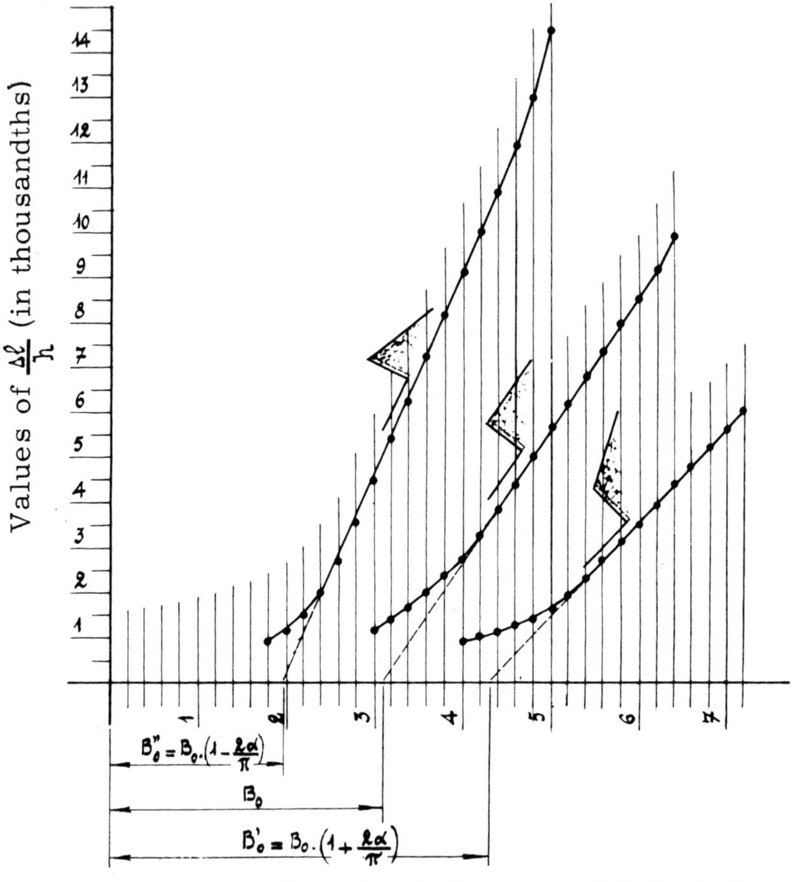

Forces pushing back the material (in kg)

Fig. 26: Elasto-plastic stage of a granular material, with horizontal or inclined top surface, loaded under conditions of translatory passive resistance: Fine sand.

sistance corresponding to a passive resistance coefficient equal to unity, in each case considered, for the material with horizontal or inclined top surface.

However, this graph does not reveal another characteristic feature which does, on the other hand, appear in Figure 27, where the same deformations determined in the tests have now been plotted, not against the forces pushing back the material, but instead against the passive resistance co-efficients as calculated and listed in the second column of the tables concerned.

This new graphical representation of the data shows all the straight lines, representing the deformations, to have the same slope during the elasto-plastic stage of equilibrium of the material. Hence the deformation $\Delta l/h$ of a mass of granular material with horizontal or inclined top surface is a linear function of the force F and therefore of the passive resistance coefficient $\mathbf{k}_p$:

$$k_p = \frac{\gamma \cdot h^2}{2} \cdot b \left(1 \pm \frac{2\,\alpha'}{\pi}\right)$$

where $\alpha = \alpha'$ in the case of a material whose top surface is inclined at the angle of repose.

This result is very important in that it completes the information already derived from the preceding test results relating to material with a horizontal top surface. It will, within the limits of the adaptation stage and the elasto-plastic equilibrium stage, enable us to establish a general rule for estimating the deformations and stresses associated with a mass of granular material having a horizontal or inclined top surface and subjected to passive resistance forces.

We see that, whatever the shape of the mass with its horizontal or inclined top surface, its deformation is represented by an expression having the following form:

$$\frac{\Delta l}{h} = \chi\,(k_p - 1) \quad {}^{*}$$

where k_p is equal to:

$$\frac{F}{\frac{\gamma \cdot h^2}{2} \cdot b}$$

---

* Experimental dertermination of the parameter χ is dealt with in Section 9.1. and in Chapter 14.

in the case of material with a horizontal top surface

$$\frac{F}{\dfrac{\gamma \cdot h^2}{2} \cdot b \cdot \left(1 + \dfrac{2\,\alpha'}{\pi}\right)}$$

in the case of material with its top surface inclined at the upper angle of repose $(+\alpha')$

$$\frac{F}{\dfrac{\gamma \cdot h^2}{2} \cdot b \cdot \left(1 - \dfrac{2\,\alpha'}{\pi}\right)}$$

in the case of material with its top surface inclined at the lower angle of repose $(-\alpha')$.

It will furthermore be seen, in Section 15.3. (Fig. 57) and in Section 16.3. (Fig. 61), that the above formula defining the deformation of a mass of material subject to translatory passive resistance is of general validity and that, under the same conditions with regard to the shape of the mass with a horizontal or inclined top surface, it is applicable also to rotational passive resistance and to toe resistance.

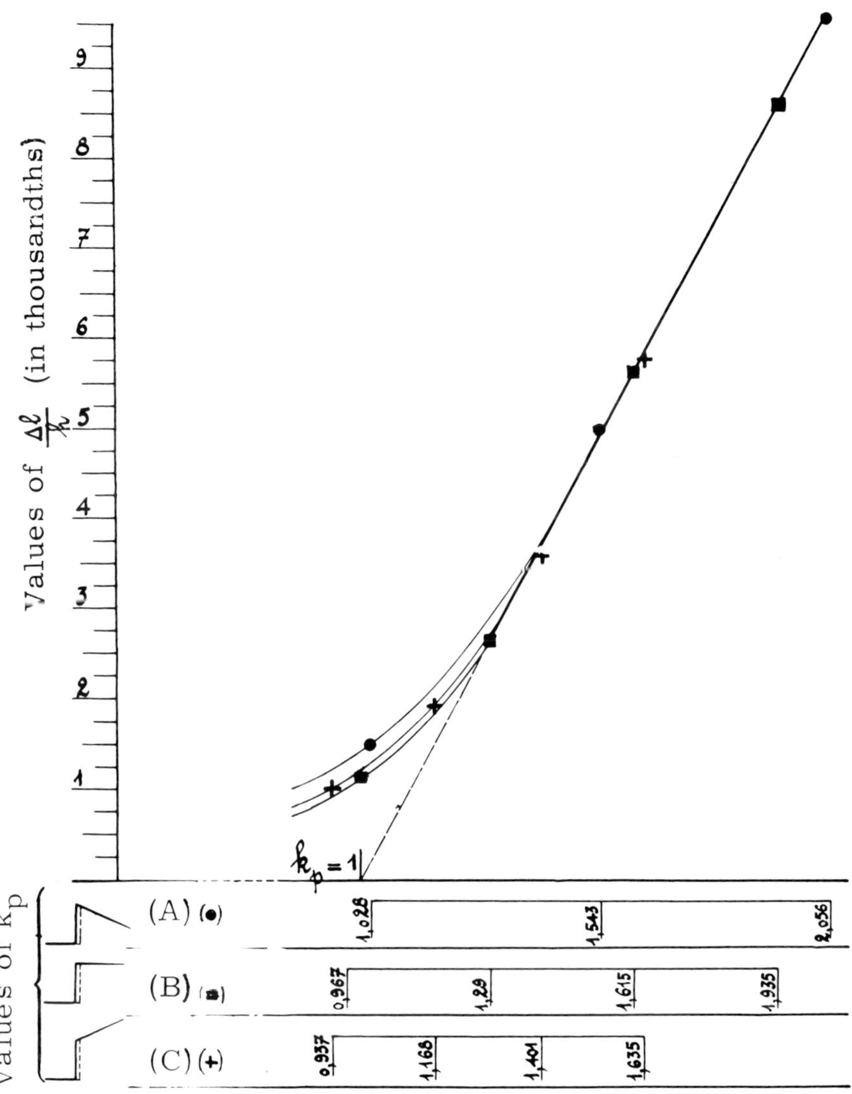

Fig. 27: Elasto-plastic stage of a granular material, with horizontal or inclined top surface:
Fine sand.
(A) top surface inclined at angle of repose $-\alpha$
(B) top surface horizontal
(C) top surface inclined at angle of repose $+\alpha$

7. Ultimate Passive Resistance Coefficients

In reporting the foregoing test results no mention was made of values of $\Delta l/h$ in the range above certain values of the passive resistance coefficient. The reason for omitting those deformations is that the results for the plastic equilibrium stage of behaviour of the material show considerable scatter, especially towards the final part of that stage.

Nevertheless, in order to give a complete account of our investigation of the deformation phenomena affecting a mass of granular material with horizontal top surface and subjected to a "pushing back" action exerted upon it by a diaphragm, we shall now give the values which we obtained for the passive resistance coefficient at failure, i.e., at the precise instant when the material ceases to offer any appreciable resistance to displacement of the diaphragm. These are extreme values which may be of some interest even though they are affected by scatter of the order of 5 to 7%.

The values $\Delta l/h$ corresponding to these ultimate passive resistance coefficients, i.e., associated with failure of the material in the manner envisaged here, are averages of those obtained in three consecutive tests continued up to the limit where the displacement of the diaphragm became abnormally large (Table 14).

Granular Materials	Ultimate Passive Resistance Coeff. k_p	$\dfrac{\Delta l}{h}$ (in thousandths)
Millet	7.9	220
Hard wheat	8.0	210
Fine sand (loose)	12.9	280
Var sand (loose)	13.4	295
Seine sand (loose)	14.2	315

Table 14

8. Confirmation, by Tests Relating to Translatory Passive Resistance, of the Thrust Coefficients for Granular Material with Inclined Top Surface

8.1. Vertical Wall

The research concerning the equilibrium of a mass of granular material at rest had led to the following results:

Thrust coefficient for material with horizontal top surface:

$$k_{a.1} = \left(\frac{\pi - 2\,\varphi_0}{\pi + 2\,\varphi_0}\right)^2$$

Thrust coefficient for material with top surface inclined at an angle $+\alpha'$ above the horizontal or an angle $-\alpha'$ below the horizontal ($\alpha' \leqslant \alpha$):

$$k_{a.2} = k_{a.1} \cdot \left(1 \pm \frac{2\,\alpha'}{\pi}\right)$$

$$= \left(\frac{\pi - 2\,\varphi_0}{\pi + 2\,\varphi_0}\right)^2 \cdot \left(1 \pm \frac{2\,\alpha'}{\pi}\right)$$

Translatory passive resistance coefficient for material with horizontal top surface:

$$k_{p.1} = k_{a.1} \cdot \left(\frac{\pi + 2\,\varphi_0}{\pi - 2\,\varphi_0}\right)^2$$

$$= \left(\frac{\pi - 2\,\varphi_0}{\pi + 2\,\varphi_0}\right)^2 \cdot \left(\frac{\pi + 2\,\varphi_0}{\pi - 2\,\varphi_0}\right)^2$$

$$= 1$$

Translatory passive resistance coefficient for material with top surface inclined at an angle $\pm\alpha'$ ($\alpha' \leqslant \alpha$):

$$k_{p.2} = k_{p.1} \cdot \left(1 \pm \frac{2\,\alpha'}{\pi}\right)$$

$$= 1 \pm \frac{2\,\alpha'}{\pi}$$

The above-mentioned passive resistance coefficients $k_{p.1}$ and $k_{p.2}$ are accurately confirmed by the experiments concerned with the passive resistance of retained granular materials, since the straight line representing the deformations occurring in the elasto-plastic stage intersects the axis of abscissae at a point corresponding to the value of the relative minimum passive resistance (for zero displacement), namely:

$$B_0 = \frac{\gamma \cdot h^2}{2} \cdot b$$

in the case of material with horizontal top surface, or

$$B_0' = \frac{\gamma \cdot h^2}{2} \cdot b \cdot \left(1 + \frac{2\,\alpha'}{\pi}\right)$$

in the case of material with top surface inclined at the upper angle of repose $(+\alpha')$, or

$$B_0'' = \frac{\gamma \cdot h^2}{2} \cdot b \cdot \left(1 - \frac{2\,\alpha'}{\pi}\right)$$

in the case of material with top surface inclined at the lower angle of repose $(-\alpha')$.

It thus happens that we here have confirmation of the characteristic general thrust coefficient: $1 \pm 2\alpha'/\pi$ on passing from material with horizontal top surface to material whose top surface is inclined at an angle $\pm\alpha'$ $(\alpha' \leqslant \alpha)$. Having regard to the present test results, this confirmation extends also to the passive resistance coefficient which had already been established (1), (2).

8.2. Sloping Wall

In the case of a wall inclined at an angle i in relation to the horizontal our research relating to the equilibrium of a mass of granular material at rest yielded the following results.

Thrust coefficient for material with horizontal top surface:

$$k_{a.3} = \left(\frac{\pi - 2\,\varphi_0}{\pi + 2\,\varphi_0}\right)^2 \cdot \left(\frac{i - \varphi_0}{\dfrac{\pi}{2} - \varphi_0}\right)$$

Translatory passive resistance coefficient for material with horizontal top surface:

$$k_{p.3} = k_{a.3} \cdot \left(\frac{\pi + 2\,\varphi_0}{\pi - 2\,\varphi_0}\right)^2$$

$$= \left(\frac{\pi - 2\,\varphi_0}{\pi + 2\,\varphi_0}\right)^2 \cdot \left(\frac{i - \varphi_0}{\dfrac{\pi}{2} - \varphi_0}\right) \cdot \left(\frac{\pi + 2\,\varphi_0}{\pi - 2\,\varphi_0}\right)^2$$

$$= \frac{i - \varphi_0}{\dfrac{\pi}{2} - \varphi_0}$$

From these two expressions for the thrust and the passive resistance relating to the same case of a mass of material with horizontal top surface and retained by a diaphragm inclined at an angle i it emerges that the test results for granular material developing *passive resistance* enable us to obtain confirmation of the value of the thrust coefficient $k_{a.3}$, for if these results confirm the value of the passive resistance coefficient $k_{p.3}$, the correctness of the value of $k_{a.3}$ is automatically deduced therefrom:

$$k_{a.3} = \frac{k_{p.3}}{\left(\dfrac{\pi + 2\,\varphi_0}{\pi - 2\,\varphi_0}\right)^2}$$

$$= k_{p.3} \cdot \left(\frac{\pi - 2\,\varphi_0}{\pi + 2\,\varphi_0}\right)^2$$

This confirmation is obtained from the tests reported below, which were carried out with fine sand, using a diaphragm 0.22 m in height and inclined at angles of 10° and 17° in relation to the vertical, so as to have a batter or, alternatively, an overhang. Thus the respective values of i (in relation to the horizontal) are as follows:

Diaphragm with batter:　　$i=100°$　(for 10°)
　　　　　　　　　　　　$i=107°$　(for 17°)

Diaphragm with overhang:　$i=\ \ 80°$　(for 10°)
　　　　　　　　　　　　$i=\ \ 73°$　(for 17°)

8.2.1. Diaphragm with 17° Overhang

Inclination of the diaphragm in relation to the horizontal:

$$i = 90° + 17° = 107°$$

Minimum translatory passive resistance, for zero displacement:

$$B_0 = \frac{1\,380 \times \overline{0.22}^2}{2} \times 0.20 \times \frac{107° - 33°\,40'}{90° - 33°\,40'} = 8.660 \text{ kg}$$

Test results are given in Table 16 and Figure 29.

Forces Pushing Back Material (in kg)	Values of $\triangle\,l$ (in 1/100 mm)
1	1.8
2	4.6
3	7.1
4	11.2
5	16.1
6	22.9
7	32.0
8	42.1
9	54.0
10	68.2
11	82.1
12	104.0
13	134.0
14	162.1
15	192.0
16	219.1
17	246.0
18	287.1

Table 15

Note:

Section 8 is more particularly concerned with seeking fresh confirmation
of the thrust coefficients and passive resistance coefficients dealt with in
Volume I. Figures 28–31 should be distinguished from the graphs specifically relating to the investigation of the deformations of a granular material
developing passive resistance, which forms the subject of the present
Volume II.

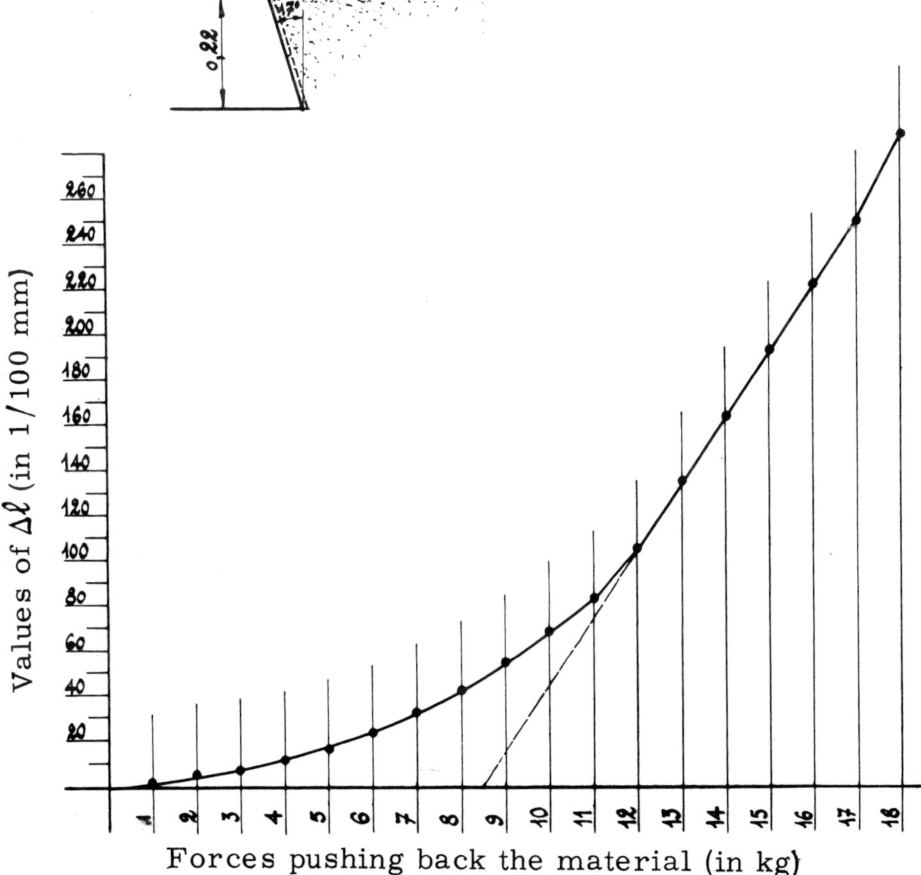

Fig. 28: Diaphragm with 17° overhang: Fine sand.
Graphical presentation of Table 15.

8.2.2. Diaphragm with 10° Overhang

Inclination of the diaphragm in relation to the horizontal:

$$i = 90° + 10° = 100°$$

Relative minimum translatory passive resistance, for zero displacement:

$$B_0 = \frac{1\,380 \times \overline{0.22}^2}{2} \times 0.20 \times \frac{100° - 33°\,40'}{90° - 33°\,40'} = 7.860 \text{ kg}$$

Test results are given in Table 15 and Figure 29.

Forces Pushing Back Material (in kg)	Values of $\triangle\,1$ (in 1/100 mm)
1	1.9
2	5.0
3	10.0
4	15.2
5	21.8
6	31.0
7	42.1
8	52.1
9	66.0
10	81.0
11	97.1
12	114.1
13	140.9
14	169.8
15	198.0
16	227.0
17	255.2
18	299.3

Table 16

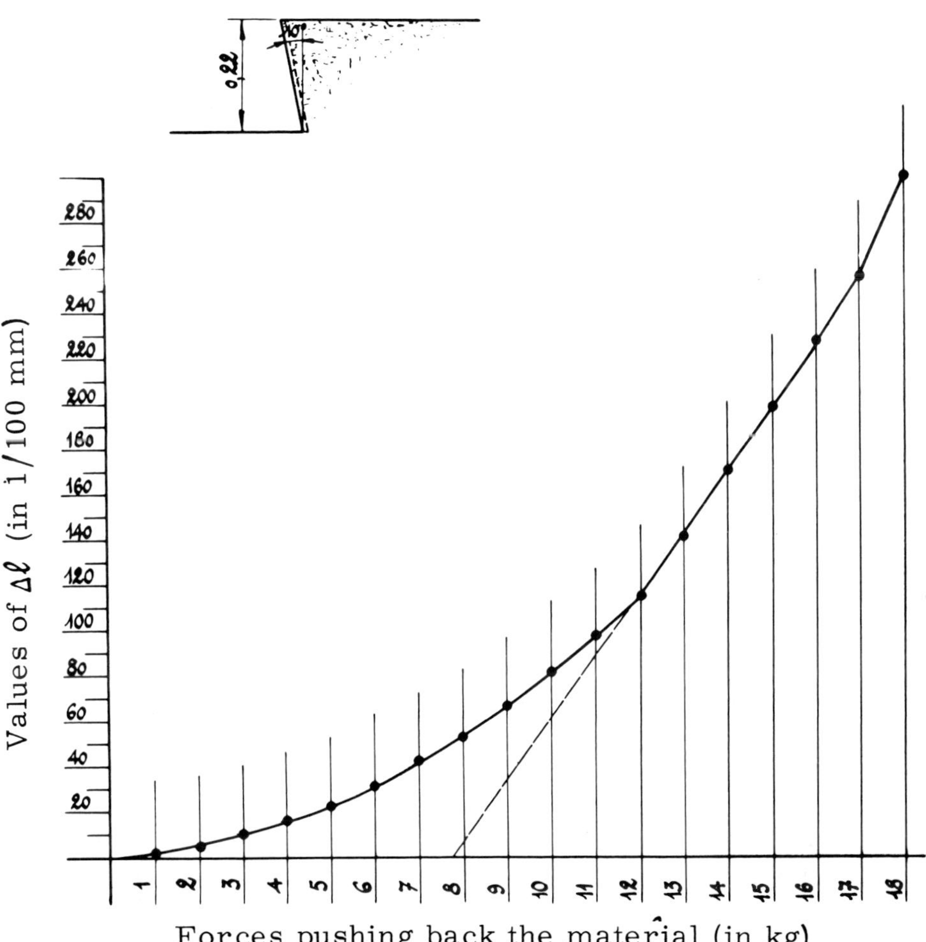

Fig. 29: Diaphragm with 10° overhang: Fine sand.
Graphical presentation of Table 16.

8.2.3. Diaphragm with 17° Batter

Inclination of the diaphragm in relation to the horizontal:

$$i = 90° - 17° = 73°$$

Relative minimum translatory passive resistance, for zero displacement:

$$\mathbf{B_0} = \frac{1\ 380 \times \overline{0.22}^2}{2} \times 0.20 \times \frac{73° - 33°\ 40'}{90° - 33°\ 40'} = 4.660 \text{ kg}$$

Test results are given in Table 17 and Figure 30.

Forces Pushing Back Material (in kg)	Values of $\triangle 1$ (in 1/100 mm)
1	2.2
2	6.2
3	15.0
4	26.0
5	42.9
6	67.0
7	89.8
8	126.0
9	160.1
10	194.0
11	224.9
12	264.0
13	311.8
14	358.0
15	410.1
16	465.3
17	544.4
18	634.1

Table 17

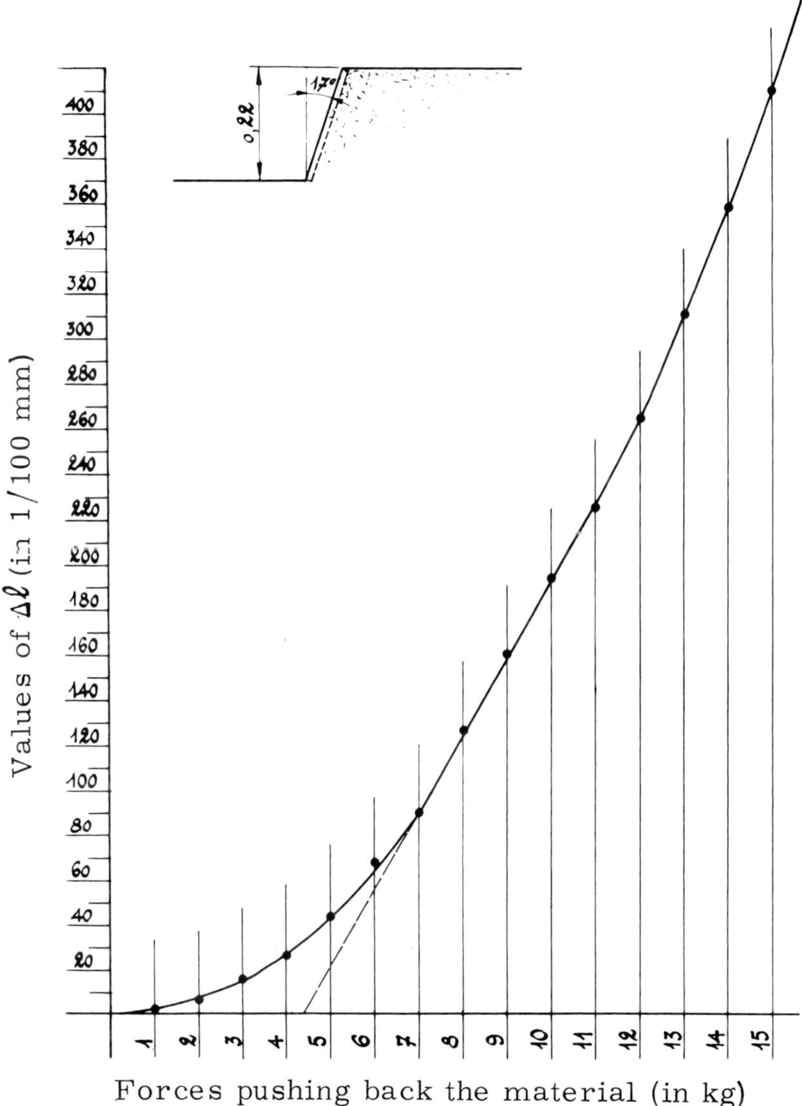

Fig. 30: Diaphragm with 17° batter: Fine sand.
Graphical presentation of Table 17.

RETAINING WALLS

8.2.4. Diaphragm with 10° Batter

Inclination of the diaphragm in relation to the horizontal:

$$i = 90° - 10° = 80°$$

Relative minimum translatory passive resistance, for zero displacement:

$$B_0 = \frac{1\,380 \times \overline{0.22}^2}{2} \times 0.20 \times \frac{80° - 33°\,40'}{90° - 33°\,40'} \cong 5.500 \text{ kg}$$

Test results are given in Table 18 and Figure 31.

Forces Pushing Back Material (in kg)	Values of $\triangle l$ (in 1/100 mm)
1	2.0
2	5.2
3	11.6
4	19.0
5	30.0
6	45.9
7	66.0
8	94.8
9	126.0
10	158.0
11	189.9
12	222.0
13	254.8
14	294.1
15	338.0
16	384.2
17	448.6
18	524.7

Table 18

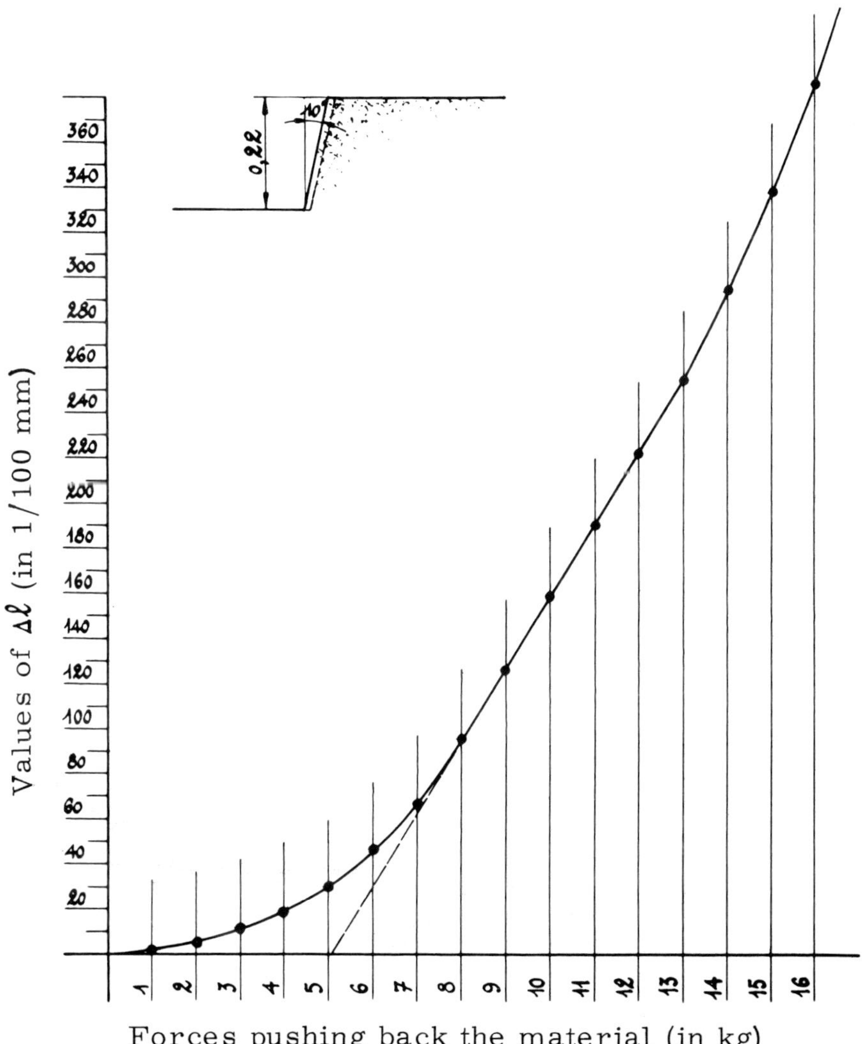

Fig. 31: Diaphragm with 10° batter: Fine sand.
Graphical presentation of Table 18.

8.3. Conclusions

An inspection of Figures 28 to 31 shows that for a mass of granular material developing passive resistance with respect to a retaining diaphragm which is sloped with an overhang or with a batter the characteristic stages of the deformation of the material are the same as those described in the case of a vertical diaphragm (Section 6.1.).

It has been shown in Section 6.1.2. that the straight line representing the deformations in the elasto-plastic stage intersect the axis of abscissae at a point which corresponds to the value of the relative minimum passive resistance for zero displacement.

In the particular case of tests performed with inclined diaphragms it is to be noted that, depending on the angle of inclination thereof, the value B_0 defined by the intersection point of the said straight line with the axis of abscissae is very close to (but slightly short of) that calculated by the formula for the case of a diaphragm with overhang of $17°$ and $10°$, whereas in the case of the diaphragm with batter of $17°$ and $10°$ the measured value of B_0 is likewise smaller than the calculated value, but now the difference is larger.

Hence it follows that in all cases the value calculated for the minimum passive resistance by means of the formula in Section 8.2. is very close to that obtained from the graphs. The same is true of the calculated thrust, the difference (in excess) being approximately:

$$2 \ \% \text{ for the diaphragm with } 17° \text{ overhang}$$
$$1.5\% \text{ for the diaphragm with } 10° \text{ overhang}$$
$$7.5\% \text{ for the diaphragm with } 10° \text{ batter}$$
$$9 \ \% \text{ for the diaphragm with } 17° \text{ batter}$$

These results therefore provide confirmation that the values of the maximum thrust coefficient and minimum passive resistance coefficient for conditions "at rest" are on the safe side by an amount of about 2% in the case of a retaining wall with overhang and of up to 10% in the case of a wall with batter.

The two formulas for maximum thrust and minimum passive resistance for inclined walls can therefore be used in full confidence because they err on the side of safety.

9. Deformations Relating to the Adaptation and Elasto-Plastic Equilibrium Stages of Granular Material Developing Translatory Passive Resistance

9.1. Review of Results

The tests reported in Section 6.2. were performed with loose sands comprising three different gradings and therefore differing also in their physical and mechanical properties, more particularly their densities and angles of internal friction.

According to those tests – performed with fine sand, Var sand (a crushed stone sand) and Seine sand – the deformations $\Delta l/h$ corresponding to the passive resistance coefficients $k_p = 1$ and $k_p = 2$ are respectively of the following order of magnitude:

$$
\begin{array}{lll}
1.24 \text{ and } 9.1 & \text{thousandths} & \text{(fine sand)} \\
1.21 \text{ and } 8.8 & \text{thousandths} & \text{(crushed stone sand)} \\
1.05 \text{ and } 8.05 & \text{thousandths} & \text{(Seine sand)}
\end{array}
$$

Rowe (4) carried out tests with a retaining diaphragm 0.45 m high and 0.60 m wide which was made to push against a mass of sand 4.20 m wide, 2.70 m long and 0.90 m deep. With loose sand this investigator obtained deformations $\Delta l/h$, for $k_p = 1$ and $k_p = 2$, of the order of 3.5 and 9 thousandths, respectively.

In these circumstances, and pending – for each particular case of construction – the laboratory determination of the parameter χ (Section 6.5.), we shall adopt the two limits 3.5 (for $k_p = 1$) and 9.1 thousandths (for $k_p = 2$) for defining the envelope of the deformations as defined in Section 8.

This envelope is represented by two straight lines (Fig. 32):

- For $k_p \leqslant 1$:

$$
\frac{\Delta l}{h} = 3.5 \, k_p \quad \text{(in thousandths)}
$$

- For

$$
1 \leqslant k_p \leqslant \frac{\pi + 2 \, \varphi_0}{\pi - 2 \, \varphi_0}
$$

(and adopting, to be on the safe side, a value of 10 for the upper limit in lieu of the experimentally determined value of 9.1.):

$$\frac{\Delta l}{h} = 3.5 + 10 \, (k_p - 1) \quad \text{(in thousandths)}$$

In the case of dense (compact) sand it will be seen that for a passive resistance coefficient k_p such as:

$$1 < k_p \leqslant \left(\frac{\pi + 2 \, \varphi_0}{\pi - 2 \, \varphi_0} \right)^2$$

the above formula is still valid, subject to the application of a reduction coefficient of 0.2 (Section 10.1.), i.e.:

$$\frac{\Delta l}{h} = 0.2 \, [3.5 + 10 \, (k_p - 1)] \quad \text{(in thousandths)}$$

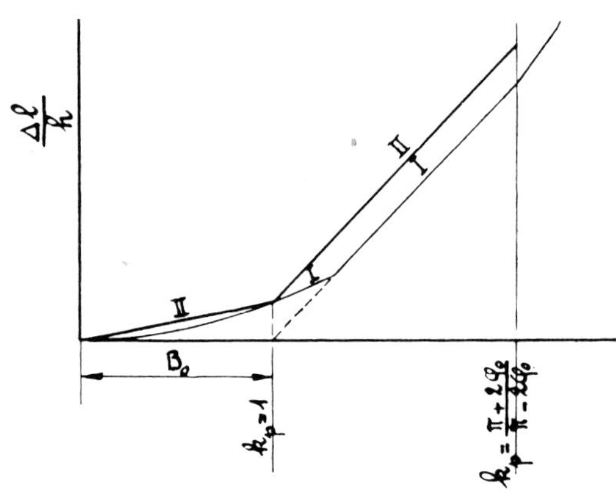

Fig. 32   (I) Deformation curve based on test results
(II) Practical enveloping curve proposed for the deformations

9.2. Special Case: Test Sands Other than Fine Sand

The tests performed with fine sand, crushed stone sand (Var sand) and Seine sand in the case of loose sand, with a 0.15 m high diaphragm, for forces corresponding to the respective values of $2 B_0$ or $0.2 \gamma h^2 b$, yielded the following values for $\Delta l/h$:

For fine sand, (Figure 17):

$$\frac{\Delta l}{h} = 9.1/1\,000$$

For crushed stone sand (Var sand), (Figure 18):

$$\frac{\Delta l}{h} = 8.8/1\ 000$$

For Seine sand (Figure 19):

$$\frac{\Delta l}{h} = 8.05/1\ 000$$

It is to be noted that these values (in thousandths) are, in relation to the value obtained for fine sand ($\Delta l/h = 9.1$), proportional to their respective densities:

$$\frac{9.1 \times 1\ 380}{1\ 420} = 8.85 \quad \text{for Var sand}$$

$$\frac{9.1 \times 1\ 380}{1\ 550} = 8.1 \quad \text{for Seine sand}$$

Similarly, for each value of B_0 for $k_p = 1$, the values of $\Delta l/h$ are approximately:

1.24 for fine sand (in accordance with Section 6.2.1.)
1.21 for Var sand (in accordance with Section 6.2.2.)
1.05 for Seine sand (in accordance with Section 6.2.3.).

It can be verified that these values conform, with good approximation, to the respective density ratios with regard to fine sand:

$$\frac{1.24 \times 1\ 380}{1\ 420} = 1.205 \quad \text{for Var sand}$$

$$\frac{1.24 \times 1\ 380}{1\ 550} = 1.10 \quad \text{for Seine sand}$$

It is therefore possible to give general expressions for the deformation of a mass of granular material having a density γ and horizontal top surface:

For $k_p \leqslant 1$

$$\frac{\Delta l}{h} = \frac{1\ 400}{\gamma} \cdot 3.5\ k_p$$

For $1 \leqslant k_p \leqslant \dfrac{\pi + 2\ \varphi_0}{\pi - 2\ \varphi_0}$

$$\frac{\Delta l}{h} = \frac{1\ 400}{\gamma} \cdot [3.5 + 10\ (k_p - 1)]$$

10. Effects of Vibration and Compaction on the Deformations of Granular Material Developing Passive Resistance

Moderate vibration of a mass of granular material retained by a diaphragm gives rise to the same deformation phenomenon in the material loaded under passive resistance conditions as in loose sand – with the same stages of adaptation, elasto-plastic equilibrium and plastic equilibrium – except that the deformations are smaller in magnitude than they are in loose material which has not been remoulded.

Applied Forces F (in kg)	Values of k_p (1)	Values of $\triangle l$ (in 1/100 mm)	Values of $\dfrac{\triangle l}{h}$ (in thousandths)
0.500	0.156	0.3	0.02
1.000	0.313	0.9	0.06
1.500	0.468	1.7	0.11
2.000	0.625	3	0.20
2.500	0.782	4.3	0.29
3.000	0.938	5.8	0.39
3.500	1.095	7.5	0.50
4.000	1.25	9.4	0.63
4.500	1.405	14.7	0.98
5.000	1.565	20.3	1.35
5.500	1.72	26	1.73
6.000	1.875	31.7	2.11
6.500	2.03	38.1	2.54
7.000	2.19	45.5	3.03
7.500	2.345	54.3	3.62
8.000	2.50	64	4.27
8.500	2.655	74.8	4.98
9.000	2.81	85.4	5.75

---

(1) $k_p = \dfrac{F}{3.200}$

Table 19

10.1. Vibrated Fine Sand

In the case of a mass of vibrated fine sand which underwent a 2% decrease in volume in consequence of vibration the following results were obtained:

Bulk density of the vibrated sand: $\gamma = 1\,407$ kg/m³
Flat vertical diaphragm: $h = 0.15$ m,
                                    $b = 0.20$ m.

Relative minimum passive resistance:

$$\mathbf{B_0} = \frac{1\,407 \times \overline{0.15}^2}{2} \times 0.20 \cong 3.200 \text{ kg}$$

Test results are given in Table 19 and Figure 33.

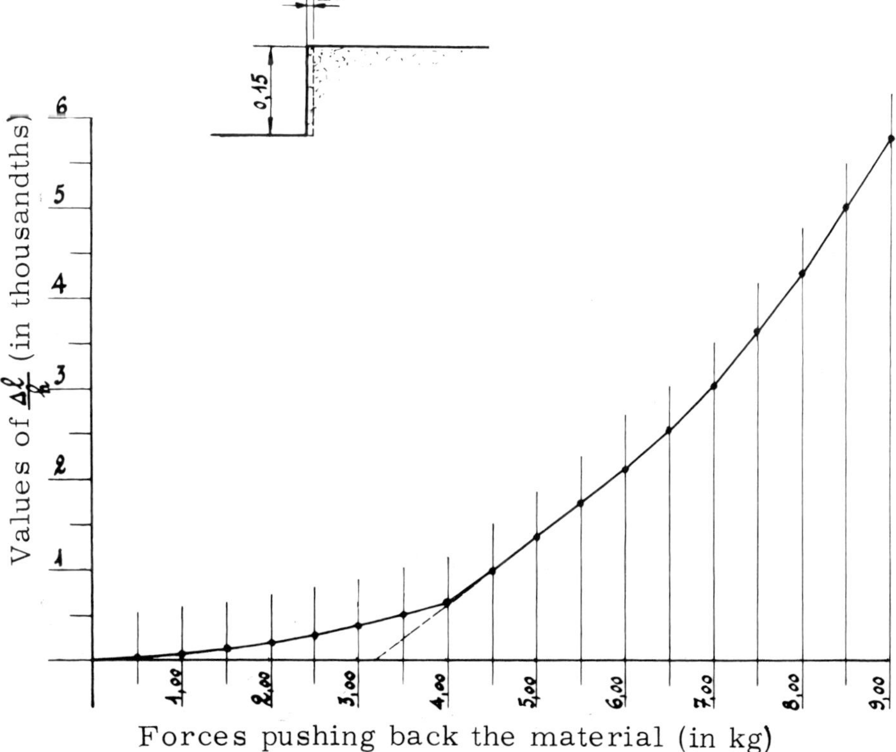

Fig. 33: Vibrated fine sand ($\gamma = 1407$ kg/m³).
Graphical presentation of Table 19.

10.2. Closely Packed Sand

In the case of a mass of sand whose particles are more closely packed as a result of stronger vibration, involving a 6% decrease in volume, the test results obtained were as follows:

Bulk density of the closely packed sand: $\gamma = 1\,463$ kg/m^3
Relative minimum passive resistance:

$$\mathbf{B_0} = \frac{1\,463 \times \overline{0.15}^2}{2} \times 0.20 \cong 3.295 \text{ kg}$$

Test results are given in Table 20 and Figure 34.

10.3. Compacted Fine Sand

Compaction profoundly alters the state of equilibrium of a mass of sand retained by a diaphragm or wall, and if the sand is heavily compacted, the deformation of the material loaded under passive resistance conditions is a linear function of the pushing-back forces applied to the diaphragm.

In the case of fine sand which has been heavily compacted to a volume reduction of 15% the following test results were obtained:

Bulk density of the compacted sand: $\gamma = 1\,590$ kg/m^3
Flat vertical diaphragm: $\qquad\qquad h = 0.15$ m,
$\qquad\qquad\qquad\qquad\qquad\quad b = 0.20$ m.

Test results are given in Table 21 and Figure 35.

Applied Forces F (in kg)	Values of k_p (1)	Values of $\triangle\, l$ (in 1/100 mm)	Values of $\dfrac{\triangle\, l}{h}$ (in thousandths)
2.000	0.61	1.5	0.1
2.500	0.76	2.2	0.15
3.000	0.91	3.0	0.2
3.500	1.06	4.2	0.28
4.000	1.21	5.5	0.37
4.500	1.36	6.75	0.45
5.000	1.52	8.3	0.55
5.500	1.67	10.2	0.68
6.000	1.82	12.1	0.8
6.500	1.97	13.8	0.92
7.000	2.12	16.6	1.1
7.500	2.28	18.7	1.25
8.000	2.43	21.1	1.4
8.500	2.58	23.2	1.55
9.000	2.73	25.5	1.7
9.500	2.88	27.8	1.86
10.000	3.03	30.1	2.0
10.500	3.19	32.2	2.14
11.000	3.34	34.8	2.32
11.500	3.49	37.0	2.46
12.000	3.64	38.9	2.6
13.000	3.95	43.3	2.88
14.000	4.25	48.7	3.25
15.000	4.55	52.4	3.5
16.000	4.86	56.9	3.8
17.000	5.16	61.6	4.11
18.000	5.46	66.1	4.4
19.000	5.76	71.9	4.8
20.000	6.07	80.9	5.4
21.000	6.37	93.1	6.2
22.000	6.68	104.9	7.0
23.000	6.98	119.8	8.0
24.000	7.28	138.1	9.2
25.000	7.59	160.6	10.7
26.000	7.89	187.6	12.5

(1) $k_p = \dfrac{F}{3.295}$

Table 20

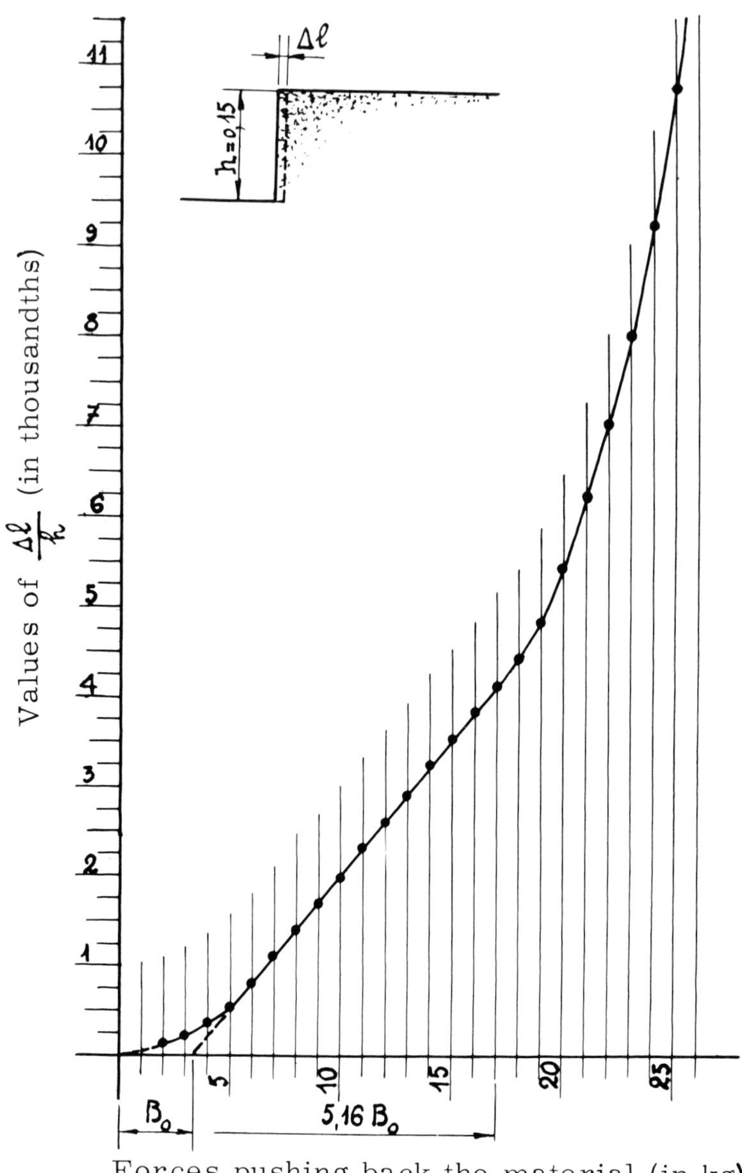

Forces pushing back the material (in kg)

Fig. 34: Closely packed fine sand ($\gamma = 1463$ kg/m³).
Graphical presentation of Table 20.

Applied Forces F (in kg)	Values of k_p (1)	Values of $\triangle$ l (in 1/100 mm)	Values of $\dfrac{\triangle\ l}{h}$ (in thousandths)
0.500	0.14	0.3	0.02
1.000	0.28	0.7	0.047
1.500	0.42	1.1	0.07
2.000	0.56	1.5	0.1
2.500	0.70	1.9	0.13
3.000	0.84	2.2	0.15
3.500	0.982	2.7	0.18
4.000	1.123	3.1	0.21
4.500	1.263	3.5	0.23
5.000	1.405	3.8	0.25
5.500	1.545	4.2	0.28
6.000	1.685	4.6	0.3
6.500	1.825	5.1	0.34
7.000	1.965	5.5	0.37
7.500	2.105	5.9	0.39
8.000	2.245	6.3	0.42
8.500	2.385	6.7	0.45
9.000	2.515	7.1	0.47
9.500	2.67	7.6	0.5
10.000	2.81	7.8	0.52
10.500	2.95	8.2	0.55
11.000	3.09	8.5	0.57
11.500	3.23	9.0	0,6
12.000	3.37	9.4	0,63
12.500	3.51	9.9	0.66

$$(1) \quad k_p = \frac{2\,F}{1\ 590 \times \overline{0.15^2} \times 0.20}$$

Table 21

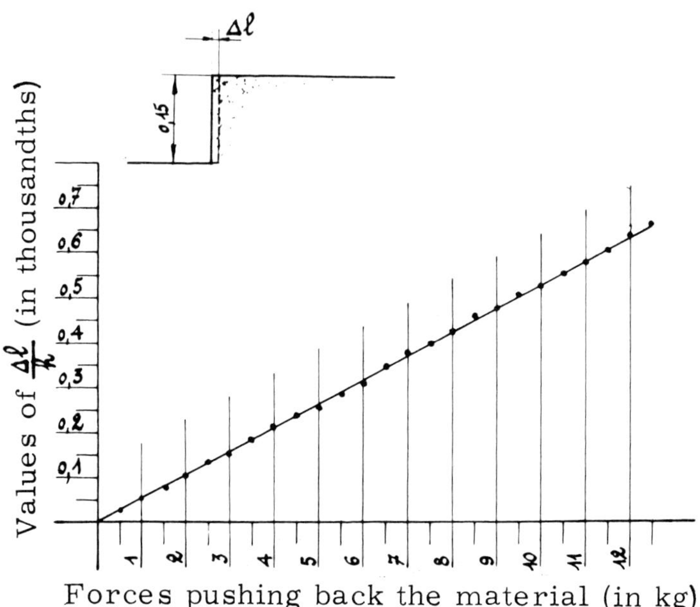

Forces pushing back the material (in kg)

Fig. 35: Compacted fine sand ($\gamma = 1587$ kg/m^3).
Graphical presentation of Table 21.

Figure 36 characteristically reveals the effect of vibration of the test material and, more particularly, the effect of heavy compaction.

It thus appears that the curve 0A in Figure 16 flattens out to become logically a straight line in that the compaction causes the sand to behave as a pseudo-elastic medium.

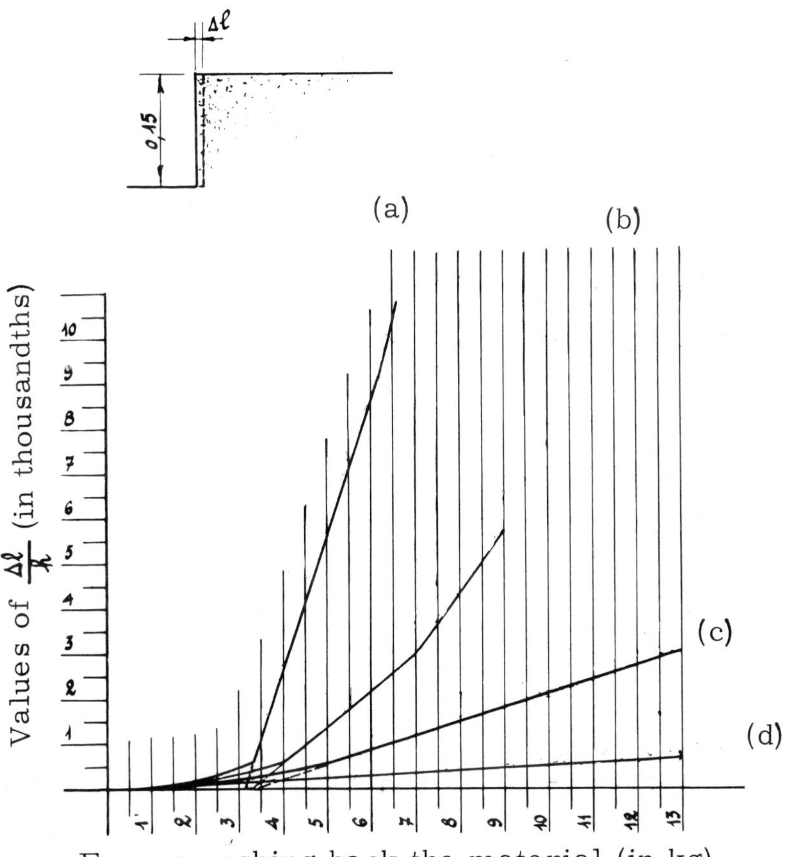

Fig. 36: Fine sand.
(a) Loose sand ($\gamma_0 = 1380$ kg/m^3)
(b) Vibrated sand ($\gamma = 1.02\,\gamma_0$)
(c) Closely packed sand ($\gamma = 1.06\,\gamma_0$)
(d) Heavily compacted sand ($\gamma = 1.15\,\gamma_0$)

10.4. Conclusions

We see in Figure 36 that the experiments carried out with sands which have been more or less energetically vibrated to a volume reduction of 6% show the same deformation phenomena as does loose sand, with the successive stages of adaptation, elasto-plastic equilibrium and plastic equilibrium, but with a very distinct decrease – in proportion as the sand undergoes closer packing of its particles, corresponding to the volume reduction of the mass – in the slope of the straight line representing the deformations in the stage of elasto-plastic equilibrium.

Another significant point to note – important in connection with defining the limits of application of the formulas for estimating deformations and stresses – is that the value of the passive resistance coefficient k_p corresponding to the limit of the elasto-plastic stage undergoes a substantial increase to something close so:

$$\left(\frac{\pi + 2\,\varphi_0}{\pi - 2\,\varphi_0}\right)^2$$

whereas we have seen in Section 6.1.3. that this value is:

$$\frac{\pi + 2\,\varphi_0}{\pi - 2\,\varphi_0}$$

in the case of loose sand.

Also, we see that between the curve representing the deformations in the case of loose sand and the curves relating to sand which has become more or less closely packed by vibration or settlement, even including natural settlement of undisturbed soil, there are an infinite number of solutions to which would correspond a whole family of straight lines representing the deformations in the stage of elasto-plastic equilibrium, ranging from k_{p2}

in the case of loose sand to k_{p2}

$$\left(\frac{\pi + 2\,\varphi_0}{\pi - 2\,\varphi_0}\right)^2$$

in the case of dense (closely packed) sand. to

$$\frac{\pi + 2\,\varphi_0}{\pi - 2\,\varphi_0}$$

From the results of the foregoing tests, as presented in Figure 36, it can be inferred that, for example, for $k_p = 2$ the value of $\Delta l/h$ is 9.1 mm for loose fine sand and 2.4 mm for vibrated fine sand with 2% volume reduction (i.e., corresponding to a ratio of the order of 4 between the respective deformations), whereas it is no more than 1 mm (i.e., a ratio of the order of 9) in the case of closely packed sand with 6% volume reduction.

Under similar conditions Rowe (4) obtained values of, respectively, 9 mm and 2 mm for $\Delta l/h$, i.e., a ratio of 4.5.

Hence it emerges that, pending the results of laboratory tests on a particular type of soil, it can conservatively (i.e., definitely on the safe side) be estimated that the deformations of undisturbed soil (that is to say: soil that has derived its compactness naturally) are only one-fifth ($=0.2$) of the deformations of loose soil.

Taking into account the passive resistance coefficients:

$$1 < k_p \leq \frac{\pi + 2\,\varphi_0}{\pi - 2\,\varphi_0}$$

in the case of loose soil, and:

$$1 < k_p \leq \left(\frac{\pi + 2\,\varphi_0}{\pi - 2\,\varphi_0}\right)^2$$

in the case of closely packed or undisturbed soil must perforce (according to the test results) be associated with the adoption of a safety factor suitably depending on the permissible amount of displacement that the designed retaining structure can undergo. A fortiori, it will be advisable to apply a higher factor of safety if the passive resistance coefficient is higher than the two foregoing values, depending on the case considered.

In the case of heavily compacted soil the deformation thereof can – in accordance with the results embodied in Figure 36 – be estimated, with a safety margin of 2, at:

$$\frac{\Delta l}{h} = 0.35\,k_p \quad \text{(in thousandths)}$$

where

$$k_p = \frac{2\,F}{\gamma \cdot h^2 \cdot b \cdot \left(1 \pm \dfrac{2\,\alpha'}{\pi}\right)}$$

in the case of translatory passive resistance.

11. Deformations in Granular Material Subjected to Surcharge

11.1. Introduction

It will be recalled that previously (1), the coefficient for maximum thrust for a mass of granular material with horizontal top surface subjected to a uniformly distributed surcharge S was defined by:

$$k_a = \left(\frac{\pi - 2\,\varphi_0}{\pi + 2\,\varphi_0}\right)^2 \cdot \left(1 + \frac{S}{\gamma h}\right)$$

and that the corresponding coefficient for relative minimum translatory passive resistance at zero displacement was:

$$k_p = k_a \cdot \left(\frac{\pi + 2\,\varphi_0}{\pi - 2\,\varphi_0}\right)^2$$

therefore

$$k_p = \left(\frac{\pi - 2\,\varphi_0}{\pi + 2\,\varphi_0}\right)^2 \cdot \left(1 + \frac{S}{\gamma h}\right) \cdot \left(\frac{\pi + 2\,\varphi_0}{\pi - 2\,\varphi_0}\right)^2$$

or

$$k_p = 1 + \frac{S}{\gamma h}$$

11.2. Experiments Concerning Translatory Passive Resistance of Material with Horizontal Top Surface and Carrying a Surcharge

It will be seen further on that in the case of a mass of granular material whose top surface is horizontal and carries a uniformly distributed surcharge and which is loaded under conditions of translatory passive resistance the phenomenon of deformation of the mass presents exactly the same stages of adaptation, elasto-plastic equilibrium and plastic equilibrium as in the case of a similar material without surcharge. This observation confirms the general character of the phenomenon already observed.

An important feature relating to the straight line representing the deformations in the elasto-plastic stage is to be noted. This line intersects the axis of abscissae – i.e., for $\Delta l/h = 0$, zero displacement of the diaphragm –

at a point which corresponds exactly to the value of the relative minimum passive resistance $B_{0.s}$ for zero displacement, calculated with the passive resistance coefficient indicated above, i.e. (per linear metre of wall):

$$B_{0.s} = \frac{\gamma h^2}{2} \cdot \left(1 + \frac{S}{\gamma h}\right)$$

The experiments relating to translatory passive resistance therefore provide fresh confirmation for the previously determined passive resistance coefficient corresponding to surcharge.

11.3. Experimental Procedure

The experiments were performed in the same way as the preceding ones relating to translatory passive resistance.

A mass of granular material was disposed against a movable diaphragm e of height h. Then a fixed diaphragm, likewise of height h, was installed above, but independent of, the movable diaphragm and the material was additionally built up to the top of this fixed diaphragm. Thus the material retained by the movable diaphragm (to which the pushing-back force was subsequently applied) was given a surcharge of height h consisting of that same material (Figure 37).

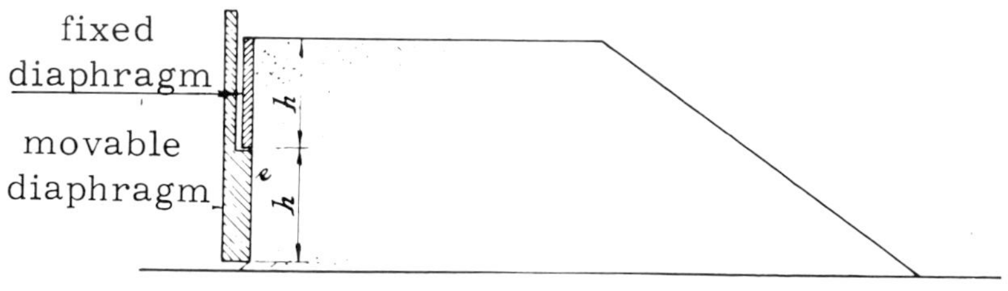

Fig. 37

The surcharge acting on the original mass was therefore:

$$S = \gamma \cdot h$$

and the passive resistance coefficient was:

$$k_p = 1 + \frac{S}{\gamma h} = 1 + \frac{\gamma h}{\gamma h}$$
$$= 2$$

with a corresponding value for the minimum passive resistance for zero displacement, which is designated by $B_{0.s}$ to distinguish it from rotational passive resistance $B_{0.r}$ and toe resistance $B_{0.cb}$ (per linear metre of wall):

$$B_{0.s} = \frac{\gamma \cdot h^2}{2} \times 2 = \gamma \cdot h^2$$

11.4. Tests

The following results were obtained with fine sand, taken as an example, for a diaphragm height $h = 0.15$ m.

Bulk density: $\qquad\qquad\qquad \gamma = 1\ 380\ \text{kg/m}^3$

Angle φ_0 $\qquad\qquad \varphi_0 \cong \alpha = 33°\ 40'$

Vertical retaining diaphragm: Height $h = 0.15$ m,
$\qquad\qquad\qquad\qquad\qquad$ Width $b = 0.20$ m.

Top surface of material is horizontal, carrying a surcharge consisting of a 0.15 m depth of fine sand. Width of wall 0.20 m. Relative minimum translatory passive resistance at zero displacement:

$$B_{0.s} = \frac{1\ 380 \times \overline{0.15^2}}{2} \times 0.20 \times 2 = 6.210\ \text{kg}$$

Test results are given in Table 22 and Figure 38.

Applied Forces F (in kg)	Values of k_p (1)	Values of $\triangle\,l$ (in 1/100 mm)	Values of $\dfrac{\triangle\,l}{h}$ (in thousandths)
1.500	0.24	0.9	0.06
2.000	0.32	1.5	0.1
2.500	0.40	2.25	0.15
3.000	0.48	3.0	0.2
3.500	0.56	3.9	0.26
4.000	0.64	4.95	0.33
4.500	0.725	6.15	0.41
5.000	0.805	7.5	0.5
5.500	0.885	9.2	0.61
6.000	0.97	11.3	0.75
6.500	1.05	13.5	0.9
7.000	1.13	16.6	1.11
7.500	1.21	21.0	1.4
8.000	1.29	24.7	1.65
8.500	1.37	30.0	2.0
9.000	1.45	36.8	2.46
9.500	1.53	43.5	2.9
10.000	1.61	49.5	3.3
10.500	1.69	56.2	3.75
11.000	1.77	62.2	4.15
11.500	1.85	69.0	4.6
12.000	1.93	75.1	5.0
12.500	2.02	82.6	5.51
13.000	2.09	88.5	5.9
13.500	2.18	96.7	6.45
14.000	2.26	106.5	7.1
14.500	2.34	118.4	7.9
15.000	2.42	133.5	8.9
16.000	2.58	166.5	11.1
17.000	2.74	212.0	13.5
18.000	2.90	246.1	16.4
19.000	3.06	293.9	19.6
20.000	3.22	352.5	23.5

$$(1)\quad k_p = \frac{F}{6.210}$$

Table 22

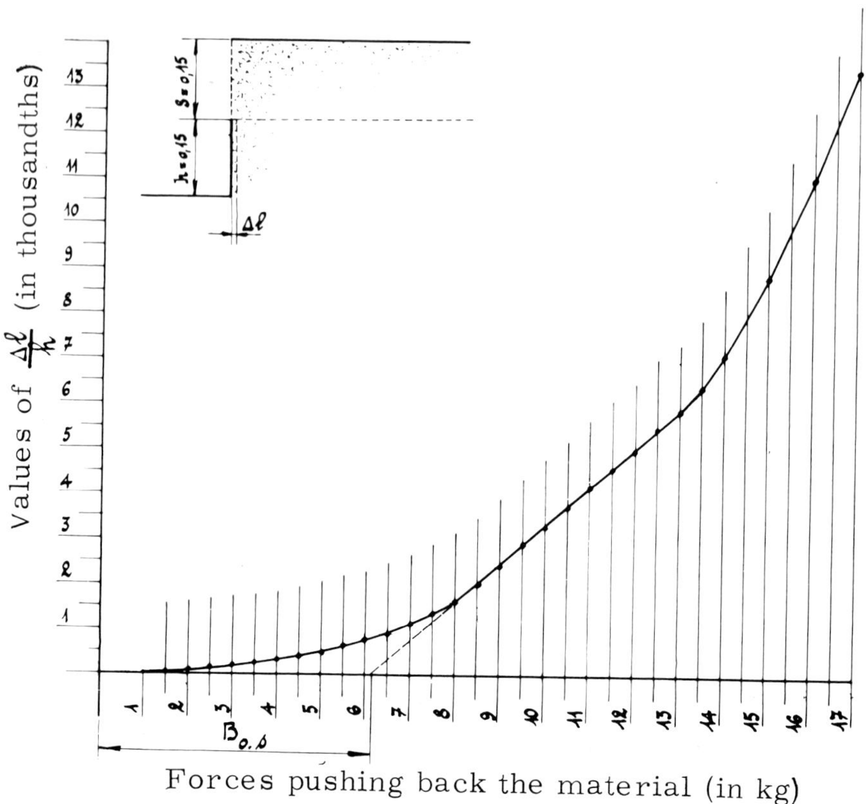

Fig. 38: Translatory passive resistance.
Material with horizontal top surface carrying uniformly distributed surcharge: Fine sand.
Graphical presentation of Table 22.

12. Dilatancy of Granular Material Developing Translatory Passive Resistance

12.1. Definition

Dilatancy designates the property of dense granular materials of undergoing an increase in volume at the instant of failure, with reduction of the coefficient φ of internal friction to its minimum value φ_0.

12.2. Experiments Relating to the Dilatancy of a Mass of Granular Material of Indefinite Extent

From the foregoing experiments in connection with the investigation of the deformations of granular material developing translatory passive resistance it was not possible to obtain much information on the phenomenon of dilatancy other than the observation of lifting (upward movement) of the top surface of the material. The commencement of this lifting coincided with the end of the stage of elasto-plastic equilibrium of the material, as an analysis of the test results showed.

Further observations revealed, however, that with a plane vertical diaphragm of width b and equal in height to 1.33 times the height of the mass of fine sand, of indefinite extent, the lifting of the horizontal top surface of the latter manifested itself as a "wave" of slightly convex shape, both transversely and longitudinally, and having an outline on plan and a cross-sectional shape as indicated in Figures 39 and 40 respectively.

This wave began to manifest itself at a distance of about 1.6 h from the diaphragm and extended, while steadily increasing in transverse width, to a distance of 1.8 h therefrom according as the granular material was pushed back more.

It was therefore necessary to pinpoint the dilatancy phenomenon which, as the test results suggested (in view of the sudden interruption of the linearity of the deformations as a function of the forces pushing back the diaphragm), manifested itself only in the range extending from the start of plastic deformations of the test material to failure corresponding to final sliding of the diaphragm.

Accordingly, fresh experiments were carried out with granular material disposed in a mass having a triangular cross-sectional shape, so as to obtain, in addition to the precise information derived from the tests, direct visual observation of the dilatancy phenomenon manifested by the test materials.

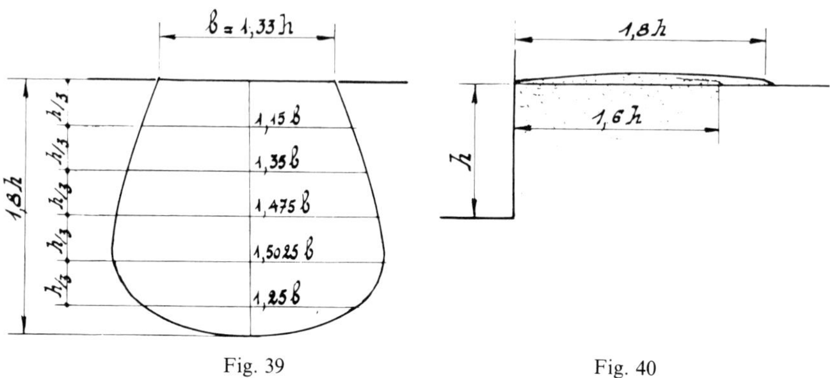

Fig. 39   Fig. 40

12.3. Experiments Relating to the Dilatancy of a Mass of Granular Material of Triangular Cross-section

Direct and precise visual observation of the dilatancy of a granular material, which was not possible for a mass having an indefinite extent, was made possible by using a mass with a triangular cross-sectional shape. The test material primarily chosen for the purpose was millet, so as to obtain a mass whose free upper surface, sloped at the angle of repose, would be highly sensitive because, as might be supposed, even the slightest displacement of the diaphragm would destroy the equilibrium at the surface, so that the grains, because of their rounded shape, would roll down the slope and thus reveal the transverse dilatancy "tendency" – i.e., the dilatancy – of the material.

The experimental investigations showed that nothing of this kind occurred at the start of the initial displacements of the diaphragm. The overall character of the deformation of the material remained exactly the same as in the case of the mass of indefinite extent, pushed back by a diaphragm, with its three stages of behaviour: adaptation, elasto-plastic equilibrium and plastic equilibrium. It was definitely established that no grain of millet moved at the free surface of the mass so long as the forces pushing back the diaphragm corresponded to only the first two stages (i.e., adaptation and elasto-plastic equilibrium), the end of which is marked by the fact that the deformations, which increase linearly in the second stage, now cease to be linear.

As soon as the deformations Δl thus ceased to increase linearly, i.e., when the third stage (plastic equilibrium) began, the grains at the surface began to detach themselves therefrom and rolled down to the foot of the slope. After this the number of grains becoming detached and rolling down the slope increased more and more as the failure limit of the mass of material was approached.

Thus we here have a direct visual observation of the deformation of the material under passive resistance conditions of loading, providing confirmation that no dilatancy phenomenon occurs during the first two stages of behaviour, namely, adaptation and elasto-plastic equilibrium.

On the other hand, throughout the stage of plastic equilibrium of the material the cross-section of the mass is continually changing as the forces applied to the diaphragm are successively increased in magnitude (Figure 41 a).

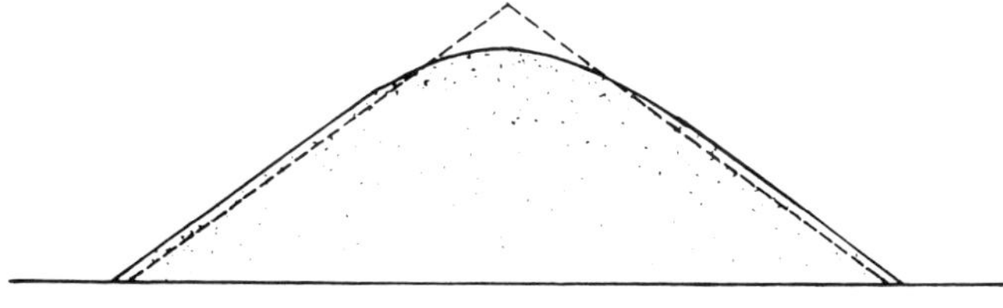

Fig. 41 a

This visual observation of the deformation of the mass is completely confirmed in the several stages of its equilibrium by results of tests performed on triangular masses consisting of millet and subsequently, as a check, on such masses consisting of fine sand from Garoupe, as reported in Sections 12.5 and 12.6.

12.4. Remark

The non-dilatancy of the triangular mass during the stages of adaptation and of elasto-plastic equilibrium is especially important in that it would, in laboratory experiments for determining the magnitude of χ to be adopted in the design of actual structures, be permissible to limit the frontal width b' of the test specimens to the width b of the diaphragm (Figure 41 b).

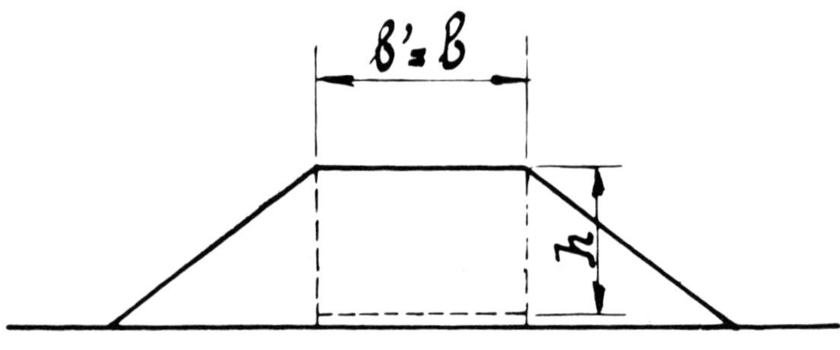

Fig. 41 b

12.5. Millet

Bulk density: $\gamma = 730 \ \text{kg/m}^3$

Angle φ_0 $\qquad \varphi_0 \cong \alpha = 28° \ 30'$

Triangular mass 0.20 m high.

Test results are given in Table 23 and Figure 42.

Applied Forces F (in kg)	Values of $\triangle$ l (in 1/100 mm)	Applied Forces F (in kg)	Values of $\triangle$ l (in 1/100 mm)
0.500	0.3	5.000	33.2
0.750	0.7	5.250	37.0
1.000	1.2	5.500	40.5
1.250	1.8	5.750	48.2
1.500	2.5	6.000	60.0
1.750	3.3	6.250	62.5
2.000	4.0	6.500	64.0
2.250	4.7	6.750	71.5
2.500	5.0	7.000	82.5
2.750	6.2	7.250	99.2
3.000	7.0	7.500	100.1
3.250	9.0	7.750	101.4
3.500	11.0	8.000	105.6
3.750	14.5	8.250	202
4.000	17.5	8.500	203
4.250	22.0	8.750	204
4.500	25.8	9.000	548
4.750	28.6		

Table 23

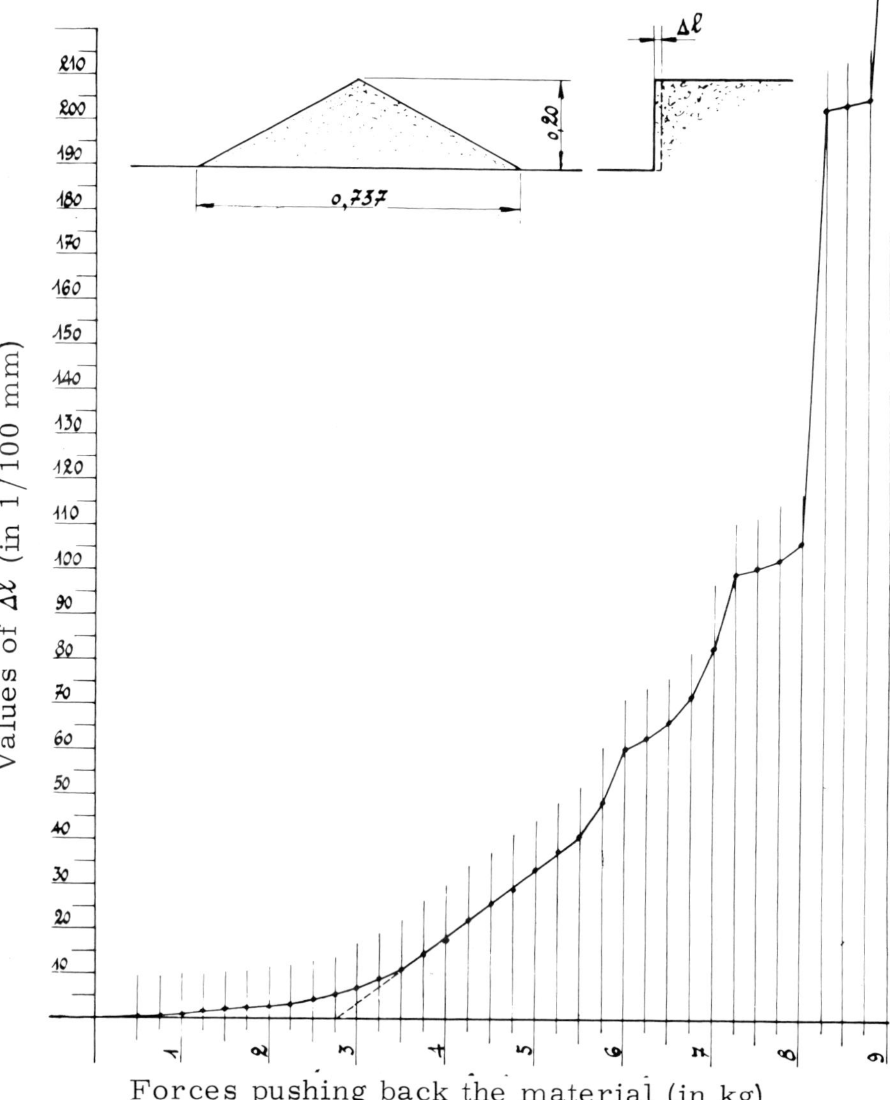

Fig. 42: Deformations Δl for millet.
Graphical presentation of Table 23.

12.6. Fine Sand

Bulk density: $\qquad\qquad \gamma = 1\,380 \text{ kg/m}^3$

Angle φ_0 $\qquad\qquad \varphi_0 \cong \alpha = 33° \, 40'$

Triangular mass 0.20 m high.

Test results are given in Table 24 and Figure 43.

Applied Forces F (in kg)	Values of $\triangle\,l$ (in 1/100 mm)	Applied Forces F (in kg)	Values of $\triangle\,l$ (in 1/100 mm)
0,500	0,15	5,000	16,5
0,750	0,3	5,250	19,4
1,000	0,65	5,500	22,8
1,250	1,0	5,750	25,3
1,500	1,4	6,000	29,0
1,750	1,7	6,250	32,0
2,000	2,2	6,500	35,0
2,250	2,8	6,750	38,0
2,500	3,4	7,000	41,0
2,750	4,1	7,250	44,0
3,000	5,0	7,500	49,1
3,250	6,3	7,750	55,0
3,500	7,5	8,000	61,0
3,750	8,5	8,250	68,1
4,000	9,8	8,500	75,0
4,250	11,0	8,750	88,5
4,500	12,0	9,000	107,5
4,750	14,0		

Table 24

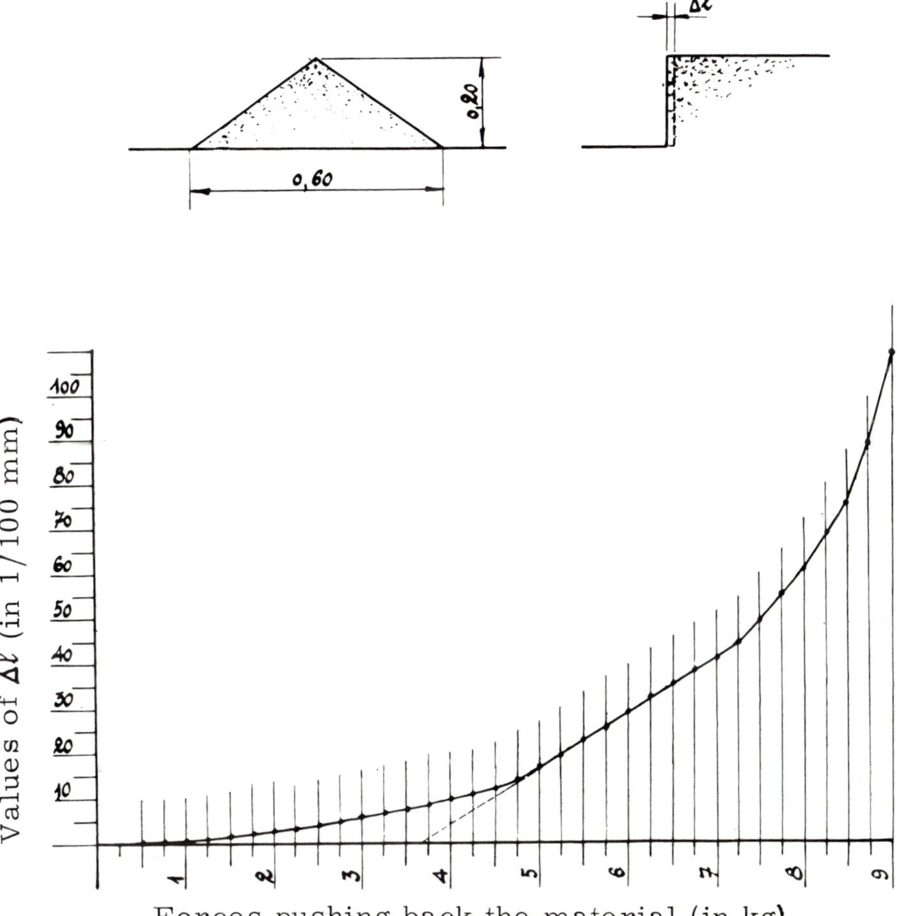

Fig. 43: Deformations Δl for fine sand.
Graphical presentation of Table 24.

13. Investigation of the Minimum Dimensions of Models

In his paper on the "language of models", Kérisel (5) called attention to the diversity of the researches undertaken with the aid of models of all sizes in connection with the investigation of active pressure and passive resistance of soils.

Having regard to the experimental results reported, we considered that it would be possible to carry out a supplementary investigation with equipment designed with due respect to the extremely high precision desired in measuring the deformations of test materials loaded under conditions of passive resistance.

After utilising the results of our first experiments, the main object of which was to study the general behaviour of the deformation phenomena relating to a mass of material with horizontal top surface and developing translatory passive resistance (Section 6), we undertook a systematic investigation with the aid of scaled-down models in order to determine the "standard model" for use in our further experimental research.

Thus, using the same descriptive terms as those in Kérisel's paper, we shall define "models comprising a diaphragm of length L and height H and having in the central part a stress-recording sensitive strip of width B_1 and height H", which diaphragm acts as a wall which retains a mass of material conceived as "a parallelepiped of width L, frontal length F and depth D" (Fig. 44, which has been reproduced from Fig. 1 in Kérisel's paper).

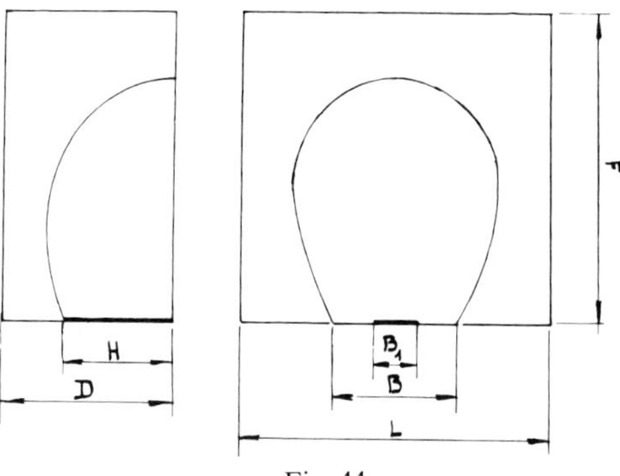

Fig. 44

13.1. Test Results

The following experiments were carried out with models having various dimensional proportions, e.g.:

Applied Forces F (in kg)	Deformations $\triangle$ l — Width L of Mass				
	1 B_1	1,5 B_1	2 B_1	2,5 B_1	3 B_1
0,600	2.2	1.8	1.0	0.7	0.4
0,800	3.6	2.8	1.9	1.3	0.9
1.000	5.0	3.8	2.6	1.9	1.3
1,200	7.2	4.5	3.5	2.2	1.7
1,400	10.0	6.9	4.6	2.8	2.1
1.600	13.8	8.8	6.0	3.5	2.6
1,800	19.5	11.0	7.5	4.5	3.2
2.000	28.5	13.5	9.5	5.5	3.9
2.200	36.0	16.6	11.9	7.1	4,7
2,400	43.1	21.1	14.0	9.0	5.6
2.600	52.5	27.3	17.0	12.1	8.9
2.800	58.9	35.2	21.1	14.9	12.7
3.000	67.0	43.1	25.3	18.4	16.7
3.200	73.6	51.9	30.2	22.6	20.4
3.400	78.5	60.3	36.9	28.1	25.3
3.600	82.4	68.2	45.5	33.1	30.0
3,800	85.5	76.1	54.1	39.3	35.2
4,000	88.6	82.1	61.2	45.1	41.0
4.200	91.2	86.8	71.8	54.2	48.2
4.400	95.1	92.1	80.0	62.8	56.2
4.600	100.2	97.5	88.8	72.6	64.5
4.800	105.5	103.5	97.6	80.4	74.2
5.000	114.1	111.0	104.1	88.6	83.4
5.200	125.0	119.5	112.2	98.8	92.2
5.400	135.2	128.0	121.5	107.1	100.5
5,600	144.8	140.1	131.2	117.5	109.5
5,800			138,5	127.5	118.6
6.000				136.5	127.6
6.200					136.5
6,400					147.5
6,600					161.0

Table 25

The results of tests performed, for example, with fine sand with a diaphragm having a width $B_1 = 0.20$ m and retaining a mass of such sand 0.15 m in height are given in Table 25 and Figure 45.

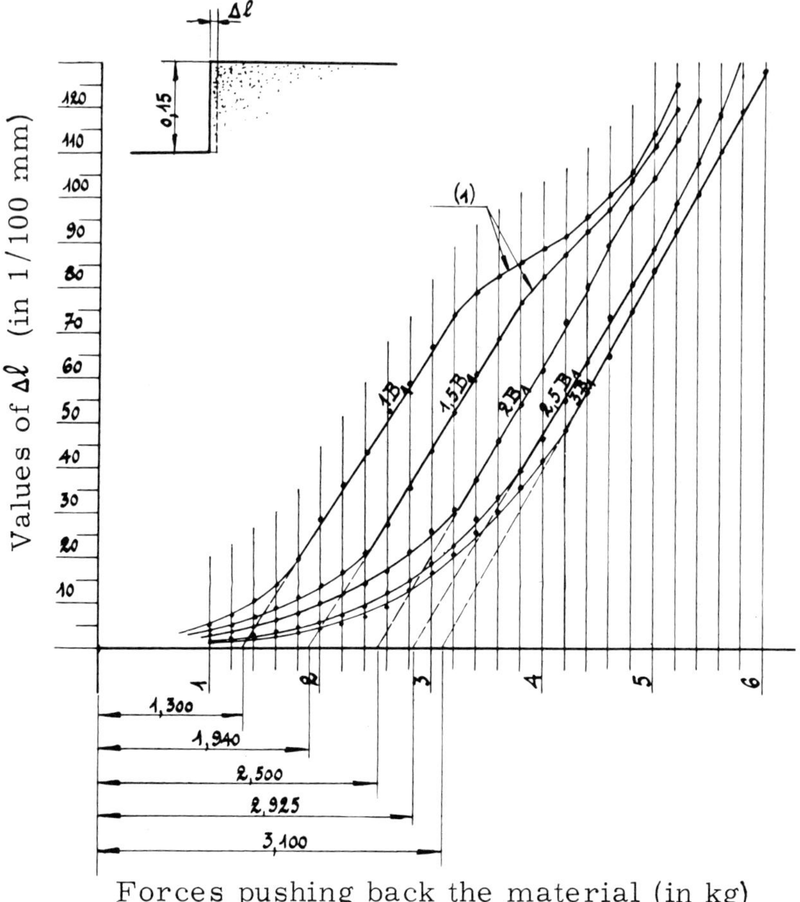

Fig. 45: Deformations Δl for fine sand.
Graphical presentation of Table 25.
(1) Anomalies in the curves are due to the "silo effect". These anomalies diminish with increasing width of the mass of sand tested.

13.2. Deformation of Material Developing Translatory Passive Resistance Across the Width L

The values of the relative minimum passive resistance, for zero displacement, corresponding to each width L of the test mass of material are as follows (these values have been scaled from the horizontal axis of the graph):

$$1{,}300 \text{ kg} \quad \text{for} \quad L = B_1 \quad = 0{,}20 \text{ m}$$

$$1{,}940 \text{ kg} \quad \text{for} \quad L = 1{,}5 \, B_1 = 0{,}30 \text{ m}$$

$$2{,}500 \text{ kg} \quad \text{for} \quad L = 2 \, B_1 \quad = 0{,}40 \text{ m}$$

$$2{,}925 \text{ kg} \quad \text{for} \quad L = 2{,}5 \, B_1 = 0{,}50 \text{ m}$$

$$3{,}100 \text{ kg} \quad \text{for} \quad L = 3 \, B_1 \quad = 0{,}60 \text{ m}$$

On plotting (Figure 46) the curve for the values of the relative minimum passive resistance, for zero displacement, we find that this resistance becomes:

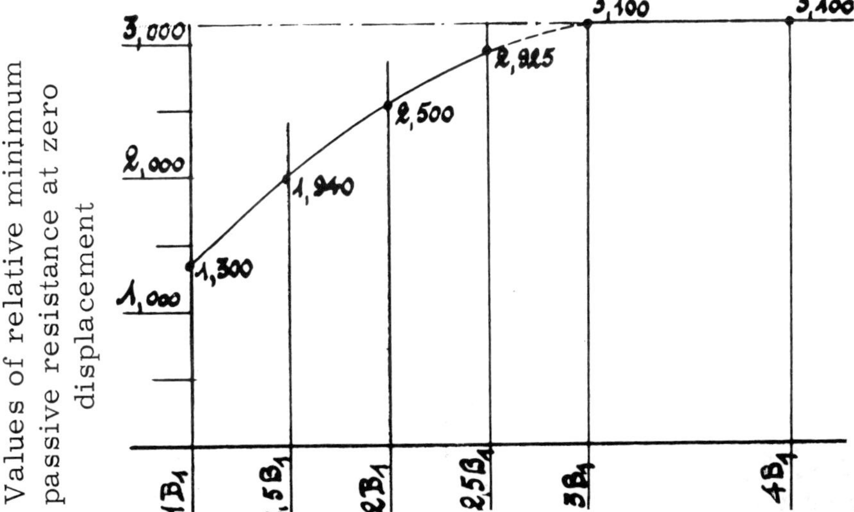

Width L of the mass as a function
of width B_1 of the diaphragm

Fig. 46

$$B_0 = \frac{1\,380 \times \overline{0.15}^2}{2} \cdot 0.20 = 3.100 \text{ kg}$$

for a value approaching $3B_1$ for the width of the mass, i.e., three times the width of the diaphragm.

This result is corroborated by the results of tests performed on a mass of material whose length is equal to three times the width of the diaphragm, because, as appears from Table 25 (represented in Figure 46), the relative minimum passive resistance, for zero displacement, is indeed equal to the calculated value, namely, 3.100 kg.

Similarly, tests with $L=4B_1$ showed the value of B_0 again to be 3.100 kg; and tests performed with material masses of the same height but of trapezoidal cross-sectional shape (i.e., freed from the lateral walls) likewise arrived at a value of 3.100 kg for B_0.

We see therefore that, for a 0.15 m high mass of fine sand, its elasto-plastic equilibrium is not disturbed, provided that the width L of the mass is at least three times the width B_1 of the diaphragm.

However, similar tests performed with a 0.20 m high diaphragm showed the minimum value of the ratio L/B_1 to be slightly larger than that corresponding to the 0.15 m high diaphragm.

It was therefore considered necessary to continue experimenting with diaphragms progressively increasing in height.

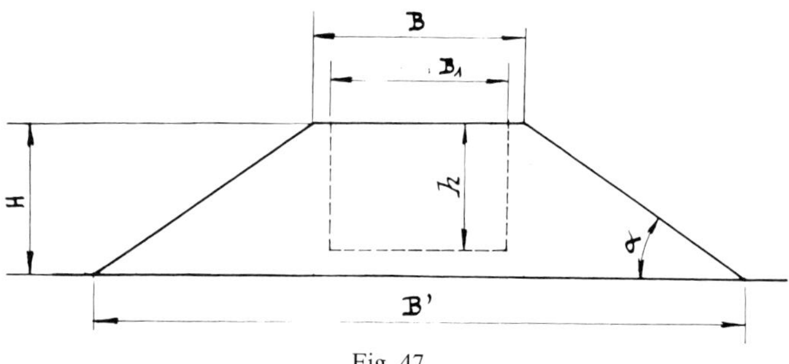

Fig. 47

In the circumstances, we preferred to rid ourselves entirely of the "silo effect" and the influence of friction on the lateral walls extending perpendicularly to the retaining diaphragm and limiting the width L of the test mass. Having regard to the information derived from the investigation of dilatancy, we simply omitted these side retaining faces and continued the experiments with material disposed in a trapezoidal mass (Figure 47), in which case a remarkably high degree of agreement between all the experiments was obtained. The top width of the trapezoidal cross-section was:

$$\mathbf{B} = 1.2\ \mathbf{B_1}$$

and the bottom width was:

$$\mathbf{B} = 1.2\ \mathbf{B_1} + 2\ \mathbf{H}\ \cotg\ \alpha$$

13.3. Note

The curves presented in Figure 45 show that, irrespective of the width L of the mass in relation to the width B_1 of the diaphragm, the same three stages of adaptation, elasto-plastic equilibrium and plastic equilibrium of the test mass are found to occur, except for a slight difference in the slope of the line representing the deformations in elasto-plastic stage.

But when the mass of material and the diaphragm have the same width $(L - B_1)$, the "silo effect" is very considerable – even if the lateral walls are polished or greased in order to obviate it – and the line corresponding to the elasto-plastic stage intersects the axis of abscissae at a value of 1.300 kg for the relative minimum passive resistance for zero displacement, i.e., very much less than the 3.100 kg thus obtained when the effect of the lateral walls is absent.

Also, it is to be noted that during the stage of plastic equilibrium the silo effect is still active in that it opposes the dilatancy of the material, thereby producing a kind of internal compressive reaction of the latter, which explains a sudden decrease in the deformations – instead of an increase as occurs in material not confined between lateral walls – as the diaphragm continues to exercise its action in pushing back the material.

On the other hand, when the ratio of material width to diaphragm width increases, the relative minimum passive resistance likewise increases in relation to the above-mentioned value and approaches the value of the relative minimum passive resistance for the material not confined between any lateral walls. The behaviour of the plastic deformations then comes closer and closer to that observed for such unconfined material.

It must therefore be borne in mind that the experiments should be performed with a mass of granular material having a width of at least about three times the width of the diaphragm or, better still, having a trapezoidal cross-sectional shape unconfined between lateral walls perpendicular to the retaining diaphragm, as in Figure 47.

13.4. Some Additional Comments

It might occur to some readers that we are rather unusual investigators in performing our experiments with masses of material having unconfined lateral and rear surfaces, so that the width L is not the same at top and at bottom level.

We wish to call our reader's attention to the tests carried out, as described in Section 13, in which the width L of the material was varied from a value equal to the width B_1 of the diaphragm to 1.5, 2, 2.5 and 3 times that width B_1. The corresponding diaphragm (Figure 46) shows that when $L=4B_1$ the value of the relative minimum passive resistance for zero displacement remains practically unchanged.

Tests performed with a mass of material having the same height, but trapezoidal in cross-section and equal in width at its top surface to the width of the diaphragm, show that the relative minimum passive resistance for zero displacement is unchanged in magnitude when the material is thus entirely unconfined by any lateral walls.

As a simple experimental precaution, and in order to be certain that the width at the top of the material will be at least equal to the width of the diaphragm, we recommend making the upper surface of the material at least 1.2 times as wide as the diaphragm.

We believe we have thus furnished experimental proof for the silo effect to which Caquot (6) drew attention in connection with tests (including even those by Terzaghi) performed on material contained in boxes, pointing out that friction on the lateral walls thereof affected – and cast suspicion on – the passive resistance values thus determined. This is indubitably revealed by Figure 46. Also, we believe we have facilitated the task of soil research laboratories in determining the function χ (as defined in Section 6.5.) in reality, because in such tests they can safely and confidently limit the volume of the test material.

We furthermore would call the reader's attention to the – in our opinion very useful – fact that henceforth experimenters will have at their disposal an accurate means of verifying the correctness of their experiments, simply by checking that the straight line representing the deformations in the elasto-plastic stage of equilibrium of the test material does indeed intersect the axis of abscissae at a point corresponding to the value of the minimum passive resistance at zero displacement.

Any test failing to satisfy this check would have to be regarded as unreliable for some reason which would have to be further investigated.

13.5. Deformations of a Material Developing Passive Resistance: Effect of the Height of the Material

The height h of the mass of material used in our experiments was varied from 10 to 15, 20, 25 and 30 cm. The corresponding values of the parameter χ as obtained experimentally were (in thousandths): 7.2, 9.1, 10.0, 10.2 and 10.5; the corresponding curve (Figure 48), which applies to loose sand, shows this parameter χ to increase only very slightly with increasing height of the retained material.

It is up to the specialised laboratories to confirm this value "in reality" and then to determine the scale effect in models. Once this has been done, it will only be necessary to extrapolate the results obtained on scaled-down models using soil samples generally of fairly small volume.

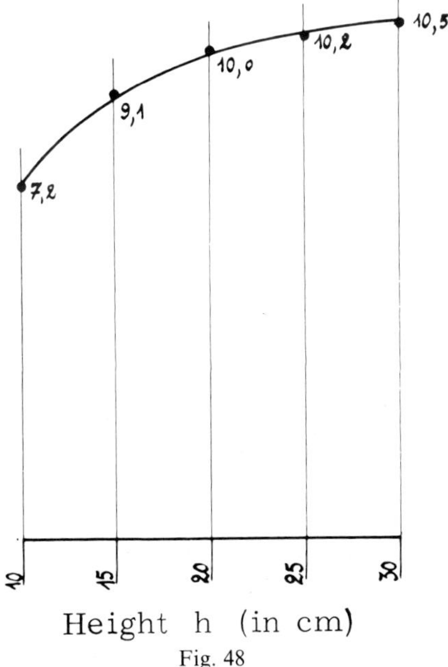

Height h (in cm)

Fig. 48

14. Practical Determination of Parameter χ in the Laboratory

From the experiments as a whole it emerges that, irrespective of the slope of the top surface of the test material (ranging from $-\alpha$ to $+\alpha$) and the surcharge applied to it, the corresponding deformation behaviour observed in the stage of elasto-plastic equilibrium of the material is always represented, as a function of the force F acting on the diaphragm and pushing back the material, by a straight line which intersects the axis of abscissae at a point which corresponds to the value of the relative minimum passive resistance B_0 for zero displacement. For a value of the passive resistance coefficient k_p corresponding to the force F exerted by the diaphragm in pushing back the material the displacement of the diaphragm is:

$$\Delta l = h \cdot \chi \, (k_p - 1)$$

where χ is a parameter to be determined for each soil specimen taken for investigation.

The task of soil testing laboratories could therefore henceforth be simplified because, for any given sample of soil, it would merely be necessary to determine χ by measuring the displacements of the diaphragm corresponding to successive values of k_p, while these values themselves would correspond to the forces successively applied to the diaphragm and ranging between approximately $1.2\ B_0$ and $2\ B_0$, for example: $1.2\ B_0$, $1.25\ B_0$, $1.30\ B_0$, $1.35\ B_0 \ldots 2\ B_0$, where B_0 is the value calculated provisionally from the general formula for the equilibrium of the granular material:

$$B_0 = \frac{\gamma \cdot h^2}{2} \cdot b \left(\frac{\pi - 2\,\varphi_0}{\pi + 2\,\varphi_0} \right)^2 \left(1 \pm \frac{2\,\alpha'}{\pi} \right) \left(\frac{\pi + 2\,\varphi_0}{\pi - 2\,\varphi_0} \right)^2$$

$$= \frac{\gamma \cdot h^2}{2} \cdot b \left(1 \pm \frac{2\,\alpha'}{\pi} \right)$$

We must plot the straight line defined by the measured displacements of the diaphragm and must extend this line to its point of intersection with the axis of abscissae k_p, which will define, on the one hand, the experimentally determined value of the relative passive resistance for zero displacement and, on the other, the value of χ to be adopted in calculating the deformations of the material: it corresponds to the value of $\Delta l/h$ for $2 B_0$ or for $k_p=2$ (Figure 51).

With a view to miniaturising the equipment that could eventually be used – in order to minimise the volume of soil needed in each sample taken for testing – it is worth noting that the results obtained do indeed conform (but with due regard to scale effect in extrapolating from the model to "the real thing") to those obtained with larger models (see Section 13). Hence it is possible to reduce the dimensions of the mass of material for testing to a height h' of 0.15 m, with a rearward length l of 0.25 m and a width b' of 0.30 m, while the diaphragm itself has a height $h=0.12$ m and a width $b=0.10$ m.

Thus the volume of the soil sample needed for the test would be of the order of 15 to 25 cubic decimetres (Figure 49).

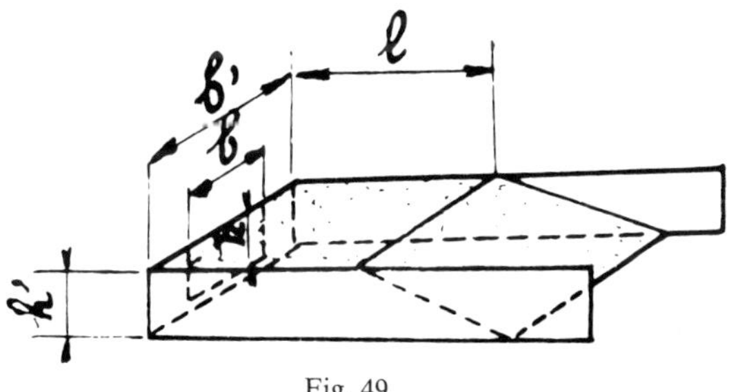

Fig. 49

However, if a larger sample, say, about 50 cubic decimetres, were available, it would be advantageous to perform the tests on an unconfined mass, i.e., without lateral walls, as shown in Figure 50: $b=0.15$ m, $b'=0.20$ m, $l=0.25$ m, $h'=0.15$ m, $h=0.12$ m.

It will be recalled that the passive resistance coefficient $k_p=1$ corresponds to the relative minimum passive resistance for zero displacement, which is equal to the value defined in the foregoing:

$$B_0 = \frac{\gamma \cdot h^2}{2} \cdot b \cdot k_p$$

in which:

$$k_p = \left(\frac{\pi - 2\,\varphi_0}{\pi + 2\,\varphi_0}\right) \cdot \left(1 \pm \frac{2\,\alpha'}{\pi}\right) \cdot \left(\frac{\pi + 2\,\varphi_0}{\pi - 2\,\varphi_0}\right)^n$$

and that the deformation corresponding to a force F acting on the diaphragm so as to push back the material is:

$$\frac{\Delta l}{h} = \chi\,(k_p - 1)$$

where:

$$k_p = \frac{F}{B_0}$$

We see that with the above formulas – for the sake of economy of design, but mobilising the passive resistance in the stage of elasto-plastic equilibrium, if the structure affected by the passive resistance reaction permits it – we can, if so desired, allow for a certain displacement Δl of the diaphragm, since we know the precise value of the corresponding passive resistance coefficient to be adopted:

$$k_p = 1 + \frac{\Delta l}{h\cdot\chi}$$

From all this it follows that the soil testing laboratories could advantageously add to their equipment an – inexpensive – apparatus based on the principle described in the foregoing. It could be used for all the tests relating to translatory passive resistance.

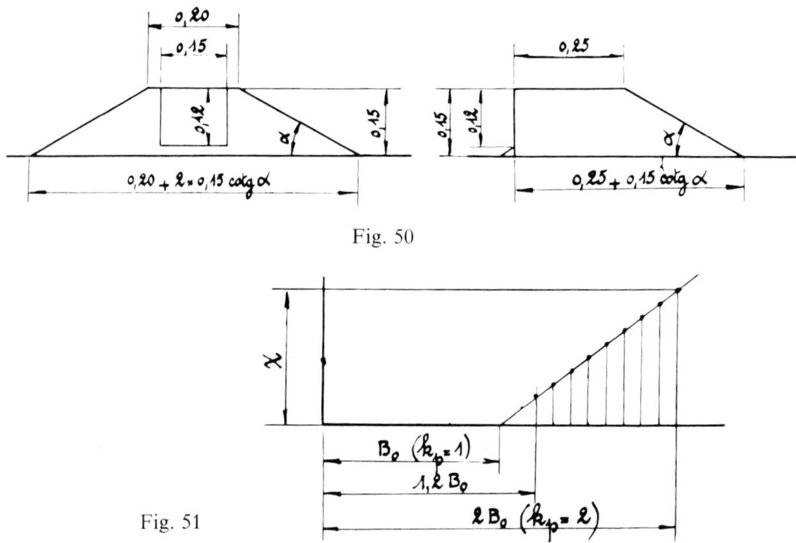

Fig. 50

Fig. 51

15. Deformations Associated with Rotational Passive Resistance

15.1. Introduction

In the experimental investigations on the translatory passive resistance developed by a mass of granular material with horizontal top surface, or top surface inclined at an angle $\pm\alpha$, it was confirmed (Section 8) that the thrust coefficient obtained previously from an interpretation of the results of our tests relating to the state of equilibrium at rest was:

$$k_a = \left(\frac{\pi - 2\,\varphi_0}{\pi + 2\,\varphi_0}\right)^2 \left(1 \pm \frac{2\,\alpha'}{\pi}\right)$$

where the positive sign $(+)$ applies to the angle α above the horizontal and the negative sign $(-)$ applies to the angle α below the horizontal.

According to our previous experiments, the rotational passive resistance coefficient – i.e., in the case where the retaining diaphragm swivels about its base – was:

$$k_p = k_a \cdot \frac{\pi + 2\,\varphi_0}{\pi - 2\,\varphi_0}$$

and therefore:

$$k_p = \left(\frac{\pi - 2\,\varphi_0}{\pi + 2\,\varphi_0}\right)^2 \cdot \left(1 \pm \frac{2\,\alpha'}{\pi}\right) \cdot \frac{\pi + 2\,\varphi_0}{\pi - 2\,\varphi_0}$$

or:

$$k_p = \frac{\pi - 2\,\varphi_0}{\pi + 2\,\varphi_0} \cdot \left(1 \pm \frac{2\,\alpha'}{\pi}\right)$$

We shall see that this coefficient is also confirmed by the experiments relating to rotational passive resistance, as described below.

15.2. Experiments Relating to the Rotational Passive Resistance of Material with Horizontal or Inclined Top Surface

Experiments relating to rotational passive resistance, i.e., when the diaphragm swivels about its base, are the easiest to perform because they do not require the use of a relatively delicately balanced mobile mounting such as the one used for measuring the deformations associated with translatory passive resistance.

15.2.1. Experimental Procedure

A mobile (pivoted) rectangular diaphragm (e) is employed, and so installed as to be able to swivel about a horizontal axis at its base, this axis being in the form of a spindle fixed on each side of the mobile diaphragm to a fixed diaphragm (e'). The mobile diaphragm has a height h and width b (Figure 52).

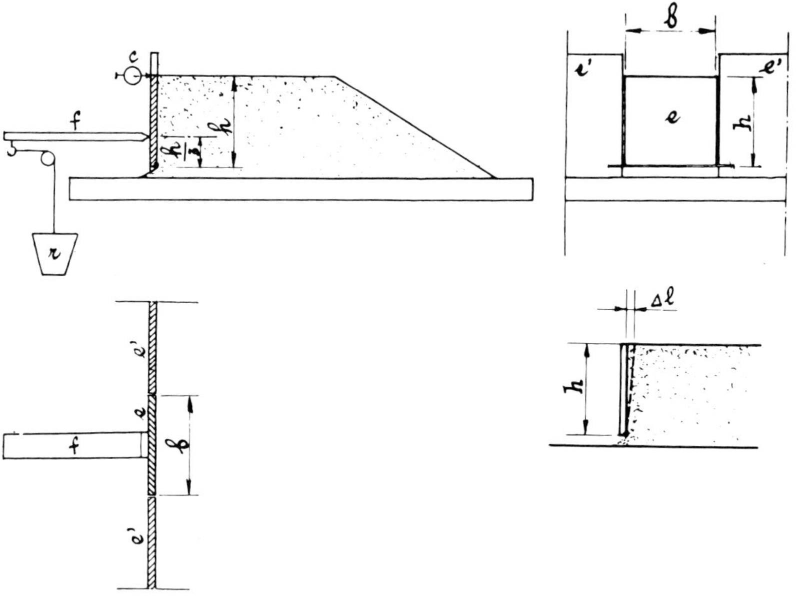

Fig. 52

A horizontally sliding thruster (f) is in contact with the mobile diaphragm at one-third of the height thereof and is moved by the pull of a string which passes round a pulley and is attached to the rear end of the thruster. The displacements Δl of the diaphragm swivelling abouts its base are measured by means of a dial gauge (c).

15.2.2. Materials with Horizontal Top Surface

The results obtained in tests performed with fine sand in a mass having a horizontal top surface are given below:

In the case of rotational passive resistance the deformation phenomena are exactly the same as those already reported in the case of translatory passive resistance, i.e., comprising the stages of adaptation, elasto-plastic equilibrium and plastic equilibrium. It will also be seen that the straight line representing the deformations in the elasto-plastic stage intersects the axis of abscissae – i.e., for zero displacement: $\Delta l/h = 0$ – at a point which corresponds to the relative minimum passive resistance B_0 at zero displacement, i.e.:

$$B_0 = \frac{\gamma \cdot h^2}{2} \cdot b \cdot \frac{\pi - 2\,\varphi_0}{\pi + 2\,\varphi_0}$$

in the case of material with horizontal top surface, and:

$$B_0 = \frac{\gamma \cdot h^2}{2} \cdot b \cdot \frac{\pi - 2\,\varphi_0}{\pi + 2\,\varphi_0} \cdot \left(1 \pm \frac{2\,\alpha'}{\pi}\right)$$

in the case of material with top surface inclined at an angle α'.

Test with Fine Sand

Bulk density: $\gamma = 1\,380 \ \text{kg/m}^3$

Angle φ_0 $\varphi_0 \cong \alpha = 33°\ 40'$

Vertical diaphragm with height $h = 0.15$ m,
width $b = 0.20$ m.

Relative minimum rotational passive resistance:

$$B_{0.r} = \frac{1\,380 \times 0.15^2}{2} \times 0.20 \times \frac{\pi - 2 \times 33°\ 40'}{\pi + 2 \times 33°\ 40'} = 1.415 \ \text{kg}$$

Test results are given in Table 26 and Figure 53.

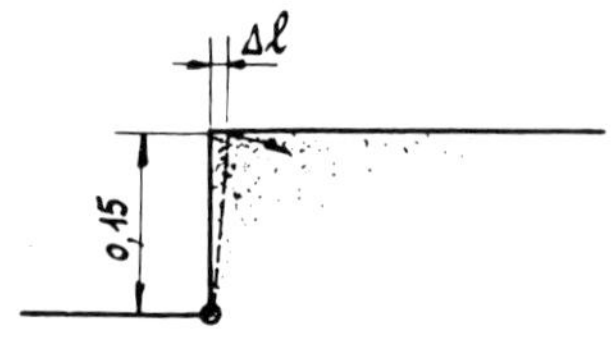

Applied Forces F (in kg)	Values of k_p (1)	Values of $\triangle\,l$ (in 1/100 mm)	Values of $\dfrac{\triangle\,l}{h}$ (in thousandths)
0.600	0.423	2.2	0.15
0.800	0.565	3.4	0.23
1.000	0.707	5.2	0.35
1.200	0.847	9.1	0.6
1.400	0.99	15.7	1.1
1.600	1.132	25.4	1.7
1.800	1.273	37.4	2.5
2.000	1.415	56.3	3.75
2.200	1.555	74.9	5.0
2.400	1.695	93.1	6.2
2.600	1.84	114.0	7.6
2.800	1.98	136.9	8.9
3.000	2.12	151.2	10.1
3.200	2.26	172.4	11.5
3.400	2.40	200.6	13.4
3.600	2.545	231.0	15.4
3.800	2.685	268.3	17.9
4.000	2.825	312.1	20.8

$$(1 \quad k_p = \frac{F}{1.415}$$

Table 26

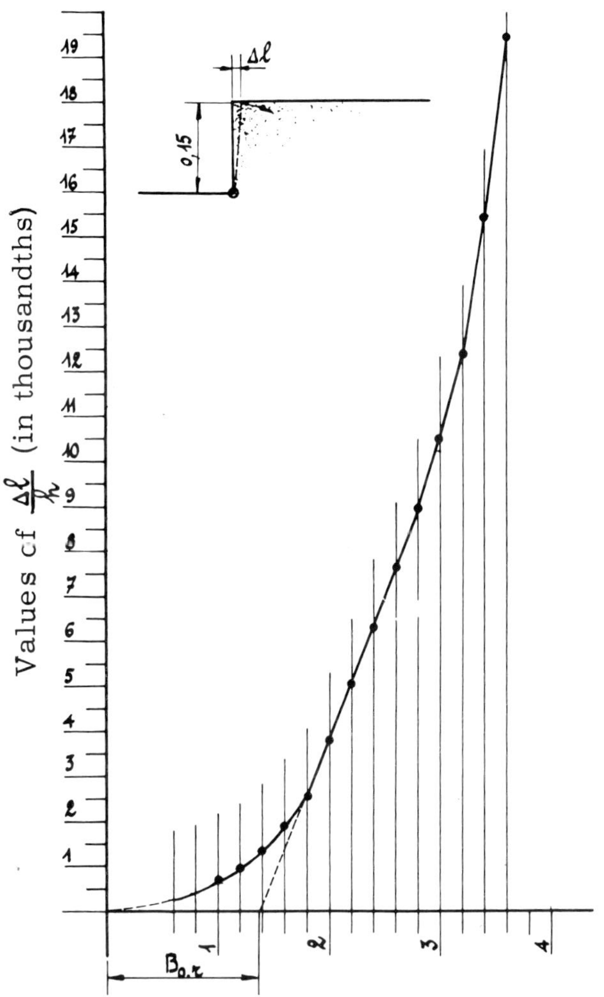

Fig. 53: Rotational passive resistance: Fine sand.
Graphical presentation of Table 26.

15.2.3. Materials with Inclined Top Surface

In the case of granular material whose top surface is inclined at an angle $\alpha'(<\alpha)$ the deformation phenomena associated with rotational passive resistance are characterised by exactly the same stages – adaptation, elasto-plastic equilibrium, plastic equilibrium – as in the case of translatory passive resistance.

The test results obtained with, for example, fine sand are summarised in Table 27.

The results are furthermore presented in two graphs. In the first of these the deformations are plotted as a function of the forces F for pushing back the material, the object being to verify – as was previously done for translatory passive resistance – that the straight line representing the deformations in the stage of elasto-plastic equilibrium of the material intersects the axis of abscissae (horizontal axis of the diagram) at a point corresponding to the value of relative minimum passive resistance for zero displacement of the diaphragm.

In the second graph the deformations are plotted against the passive resistance coefficient in order to obtain confirmation of the relationship already established in interpreting the results of the tests relating to translatory passive resistance.

According to that relationship the deformation of a mass of granular material under passive resistance loading conditions during the stage of elasto-plastic equilibrium thereof is a linear function of the passive resistance coefficient corresponding to the forces exerted on the diaphragm and striving to push back the material; this linear behaviour holds good irrespective of whether the top surface of the material is horizontal or inclined.

Note

The relative minimum passive resistance, for zero displacement, in the case of rotational passive resistance is designated by $B_{0.r}$ in order to distinguish it from the minimum translatory passive resistance designated in general by B_0.

Test with Fine Sand

Bulk density: $\qquad \gamma = 1\ 380\ \text{kg/m}^3$

Angle $\varphi_0 \qquad \varphi_0 \cong \alpha = 33°\ 40'$

Material has top surface inclined at upper angle of repose $+\alpha$.

Vertical diaphragm with height $h = 0.15$ m,
width $b = 0.20$ m.

Relative minimum rotational passive resistance:

$$\mathbf{B}'_{0.r} = \frac{1\ 380 \times \overline{0.15}^2}{2} \times 0.20 \left(1 + \frac{2 \times 33°\ 40'}{180°}\right) \cdot \frac{\pi - 2 \times 33°\ 40'}{\pi + 2 \times 33°\ 40'}$$

$$= 1.945\ \text{kg}$$

Test results are given in Table 27 and Figure 54. It can be verified from Figure 54 that the straight line representing the deformations in the stage of elasto-plastic equilibrium of the material intersects the axis of abscissae at a point corresponding to the value of the relative minimum passive resistance for zero displacement of the diaphragm.

Test with Fine Sand

Bulk density: $\qquad \gamma = 1\ 380\ \text{kg/m}^3$

Angle $\varphi_0 \qquad \varphi_0 \cong \alpha = 33°\ 40'$

Material has top surface inclined at lower angle of repose $-\alpha$.

Vertical diaphragm with height $h = 0.15$ m,
width $b = 0.20$ m.

Relative minimum rotational passive resistance:

$$\mathbf{B}''_{0.r} = \frac{1\ 380 \times \overline{0.15}^2}{2} \times 0.20 \left(1 - \frac{2 \times 33°\ 40'}{180°}\right) \cdot \frac{\pi - 2 \times 33°\ 40'}{\pi + 2 \times 33°\ 40'}$$

$$= 0.885\ \text{kg}$$

Test results are given in Table 28 and Figure 55.

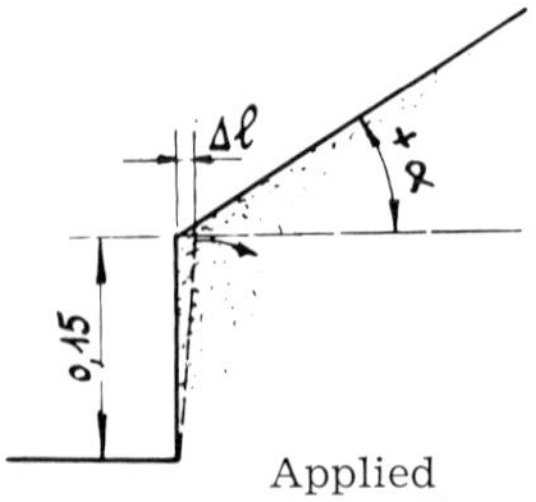

Applied Forces F (in kg)	Values of k_p (1)	Values of $\triangle l$ (in 1/100 mm)	Values of $\dfrac{\triangle l}{h}$ (in thousandths)
0.600	0.308	3.0	0.2
0.800	0.412	4.2	0.28
1.000	0.514	5.7	0.38
1.200	0.616	7.5	0.50
1.400	0.72	10.5	0.70
1.600	0.823	13.5	0.90
1.800	0.925	17.2	1.15
2.000	1.028	22.6	1.50
2.200	1.131	28.4	1.895
2.400	1.234	35.2	2.35
2.600	1.337	46.5	3.10
2.800	1.44	59.9	4.0
3.000	1.543	73.5	4.9
3.200	1.645	89.2	5.85
3.400	1.748	102.1	6.8
3.600	1.85	115.5	7.7
3.800	1.954	130.5	8.7
4.000	2.055	142.5	9.5
4.200	2.16	157.6	10.5
4.400	2.261	174.2	11.6
4.600	2.365	199.0	13.25
4.800	2.463	232.2	15.5
5.000	2.57	288.0	19.2

(1) $\quad k_p = \dfrac{F}{1.945}$

Table 27

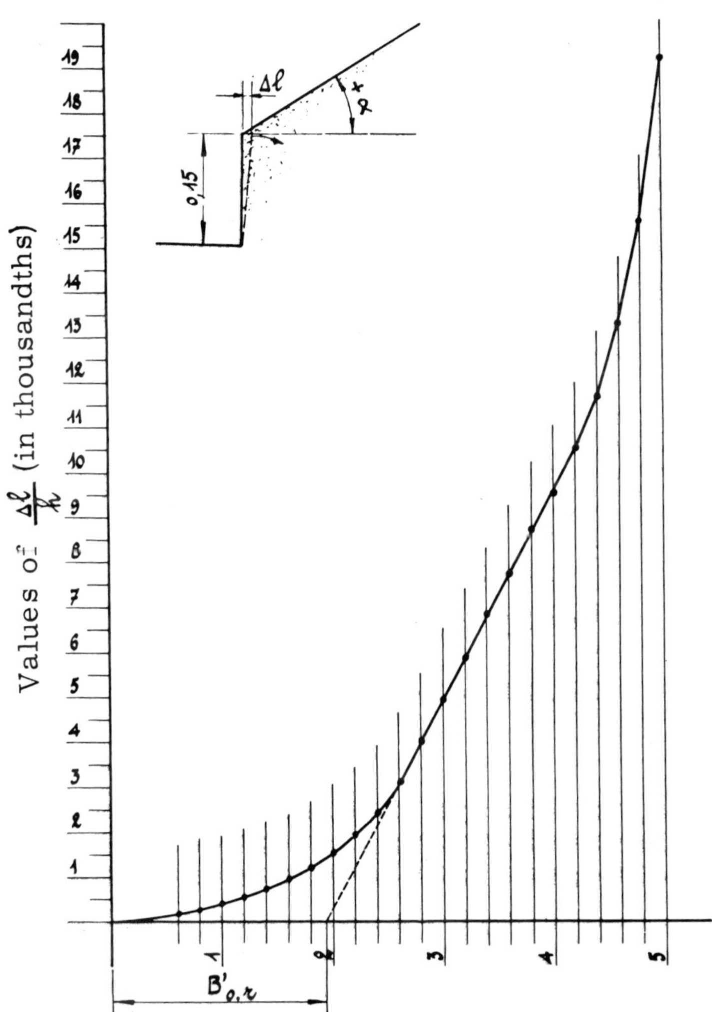

Fig. 54: Rotational passive resistance: Fine sand.
Graphical presentation of Table 27.

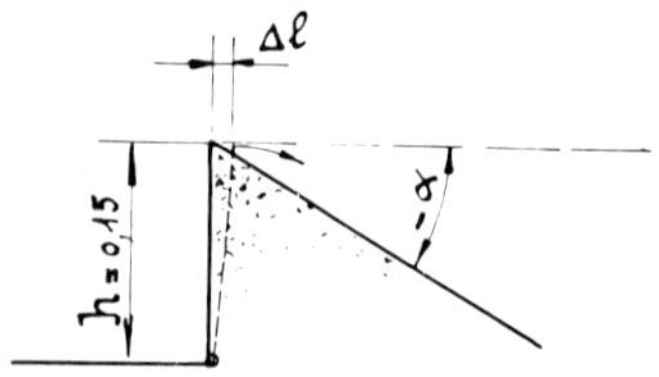

Applied Forces F (in kg)	Values of k_p (1)	Values of $\triangle$ l (in 1/100 mm)	Values of $\dfrac{\triangle l}{h}$ (in thousandths)
0.600	0.678	6.1	0.4
0.800	0.904	10.4	0.7
1.000	1.13	22.6	1.5
1.200	1.356	46.4	3.1
1.400	1.58	73.1	5.2
1.600	1.808	109.4	7.3
1.800	2.032	139.6	9.3
2.000	2.26	169.1	11.3
2.200	2.485	198.0	13.2
2.400	2.71	223.4	14.9
2.600	2.94	267.0	17.8
2.800	3.16	318.1	21.2
3.000	3.39	sliding	

$$k_p = \frac{F}{0.885}$$

Table 28

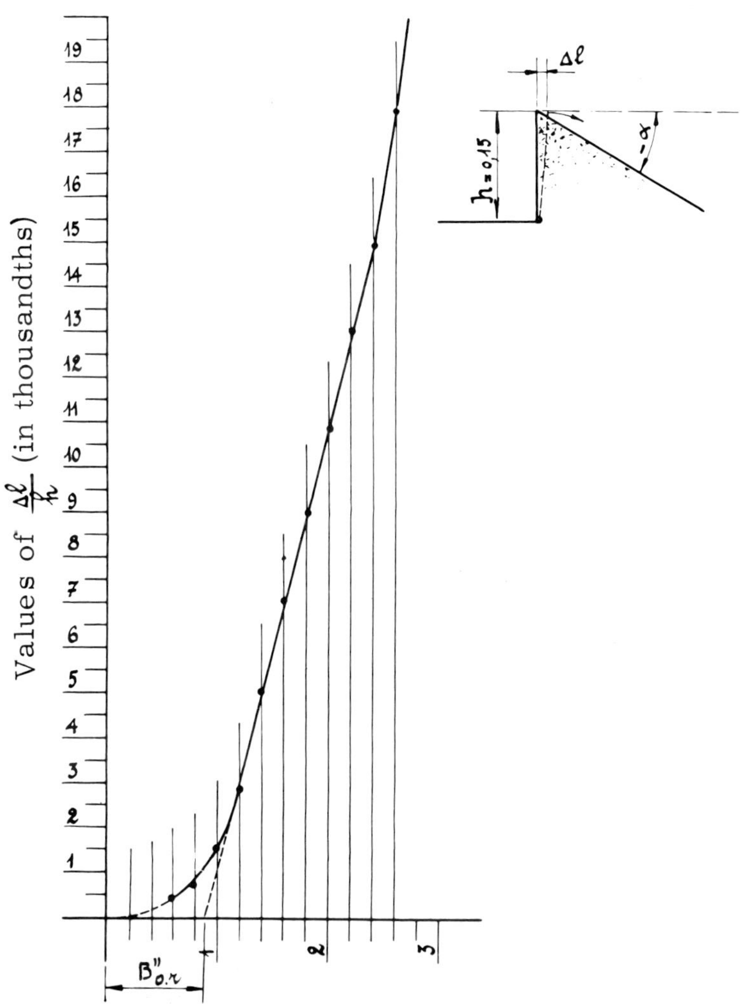

Forces pushing back the material (in kg)

Fig. 55: Rotational passive resistance: Fine sand.
Graphical presentation of Table 28.

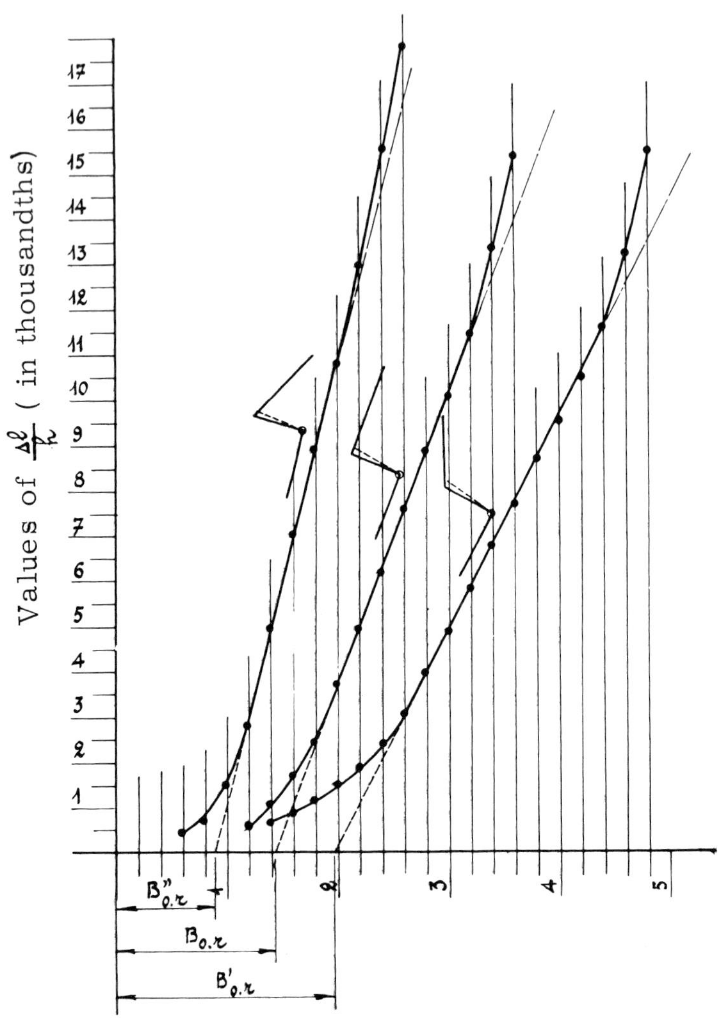

Fig. 56: Rotational passive resistance: Fine sand.
Deformations as a function of the forces pushing back the diaphragm in the case of material with horizontal or with inclined top surface.

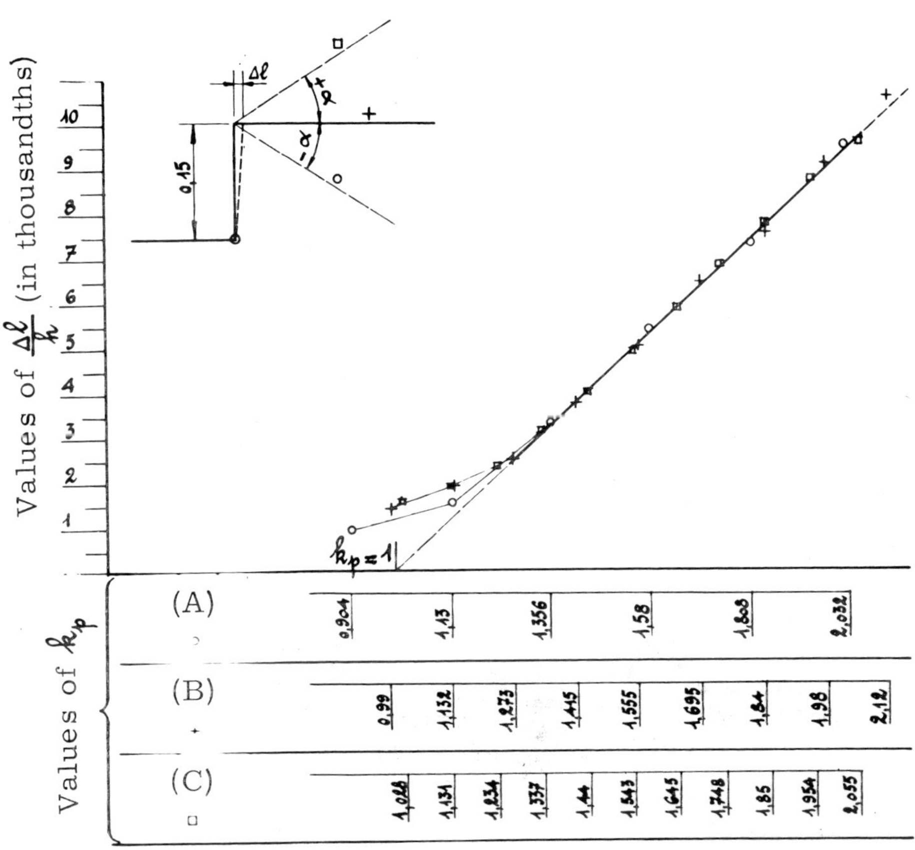

Fig. 57: Rotational passive resistance: Fine sand.

Deformations as a function of the passive resistance coefficient.

(A) top surface inclined at angle of repose $-\alpha$
(B) top surface horizontal
(C) top surface inclined at angle of repose $+\alpha$

16. Deformations Associated with Toe Resistance

16.1. Introduction

The experiments relating to what in this book is referred to as toe resistance – when the diaphragm tending to push back the material swivels about its top horizontal edge – were performed with the same experimental equipment as that used for studying rotational passive resistance phenomena. The only difference was that now the axis of rotation was at the top edge of the diaphragm.

The experiments show that with toe resistance the deformation phenomena are exactly similar to those previously observed in the cases of translatory and of rotational passive resistance respectively, with the same three stages: adaptation, elasto-plastic equilibrium and plastic equilibrium of the material.

The test results obtained with, for example, fine sand are summarised in Table 29. These results are moreover represented in three graphs (Figures 58, 59, 60).

In the first (Figure 58), in which the deformations have been plotted against the forces F pushing back the material, the value of the relative minimum passive resistance, at zero displacement of the diaphragm, can be established by determining the point of intersection of the axis of abscissae with the straight line representing the deformations in the stage of elasto-plastic equilibrium of the material.

In the second graph (Figure 59) the deformations have been plotted against the passive resistance coefficient corresponding to the value of the relative minimum passive resistance, for zero displacement, as determined from the preceding graph. This second graph provides confirmation of the general character of the relationship deduced from the interpretation of the test results relating to translatory and to rotational passive resistance, according to which – as here emerges once again for the case of toe resistance – the deformation of a mass of granular material under passive resistance loading conditions is, during the stage of elasto-plastic equilibrium of the material, a linear function of the passive resistance coefficient.

This same graph (Figure 59) also summarises the results previously obtained in the investigation of translatory and of rotational passive resistance. It provides confirmation of the constancy – for any one and the same granular material and for a given height of the retaining diaphragm – of the value of the parameter χ (Section 17), irrespective of the manner in which the passive resistance of the material is developed, namely, as translatory or as rotational passive resistance or as toe resistance.

For these three types of passive resistance in the case of material with horizontal top surface the deformations have been plotted in the third graph (Figure 60) against the force F pushing back the material. Thus the curves for these three types can readily be compared with one another in the same diagram.

It will more particularly be noted that, in relation to the curve for translatory passive resistance, the toe resistance curve is very close to that for rotational passive resistance.

Thus the ratio between the relative minimum toe resistance coefficient at zero displacement and the relative minimum rotational passive resistance coefficient is only 1.1.

Thus, in establishing a practical rule for estimating the deformations of a granular material developing passive resistance, the toe resistance can be treated as rotational passive resistance, as this will involve an over-estimation of only about 10%.

16.2. Tests with Fine Sand

Bulk density: $\qquad \gamma = 1\ 380\ kg/m^3$

Angle $\varphi_0 \qquad \varphi_0 \cong \alpha = 33° 40'$

Vertical diaphragm with height $h = 0.15$ m,
$\qquad\qquad\qquad$ width $b = 0.20$ m.

Relative minimum rotational passive resistance, at zero displacement, determined from the graph representing the test results:

$$\mathbf{B_{0.cb}} \cong 1.600\ \mathbf{kg}$$

Test results are given in Table 29 and Figure 58.

The relative minimum toe resistance, for zero displacement, is designated by $B_{0.cb}$ in order to distinguish it from the minimum translatory passive resistance designated in general by B_0 and from the minimum rotational passive resistance designated by $B_{0.r}$.

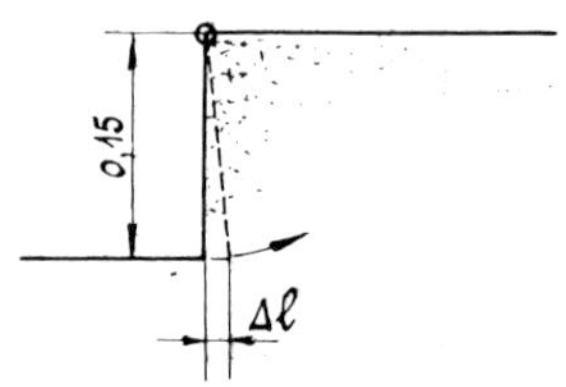

Applied Forces F (in kg)	Values of k_p (1)	Values of $\triangle l$ (in 1/100 mm)	Values of $\dfrac{\triangle l}{h}$ (in thousandths)
0.600	0.375	2.2	0.15
0.800	0.500	3.8	0.25
1.000	0.625	5.2	0.35
1.200	0.750	6.7	0.45
1.400	0.875	9.0	0.60
1.600	1.000	15.7	1.05
1.800	1.125	23.2	1.55
2.000	1.250	36.1	2.40
2.200	1.375	53.9	3.60
2.400	1.500	70.3	4.68
2.600	1.625	86.3	5.75
2.800	1.750	107.1	7.15
3.000	1.875	123.6	8.25
3.200	2.000	141.0	9.40
3.400	2.12	159.2	10.6
3.600	2.250	182.2	12.15
3.800	2.37	194.2	13.85
4.000	2.500	235.1	15.65

(1) $k_p = \dfrac{F}{1\ 600}$

Table 29

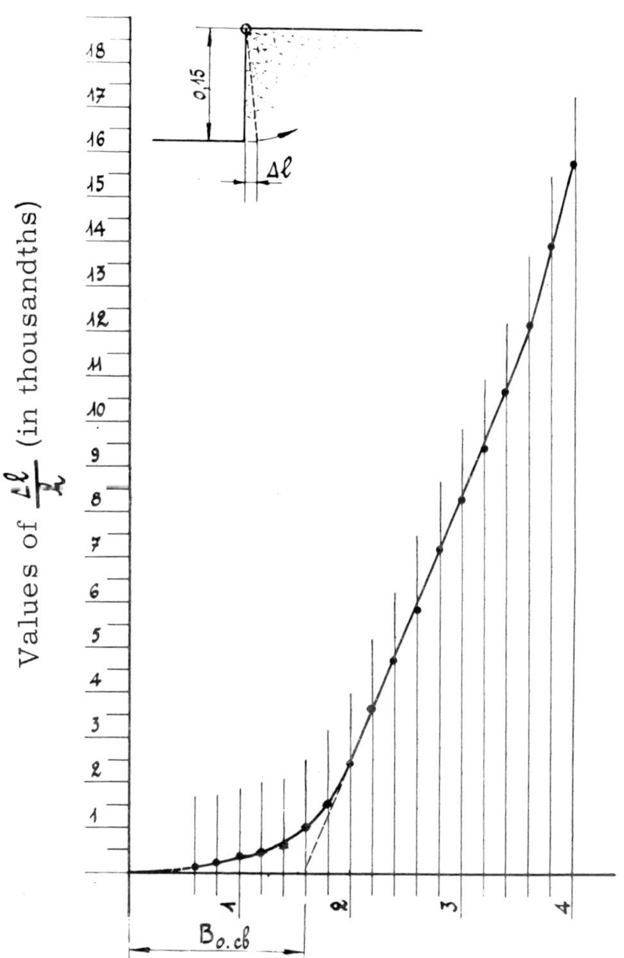

Forces pushing back the material (in kg)

Fig. 58: Toe resistance: Fine sand.
Graphical presentation of Table 29.

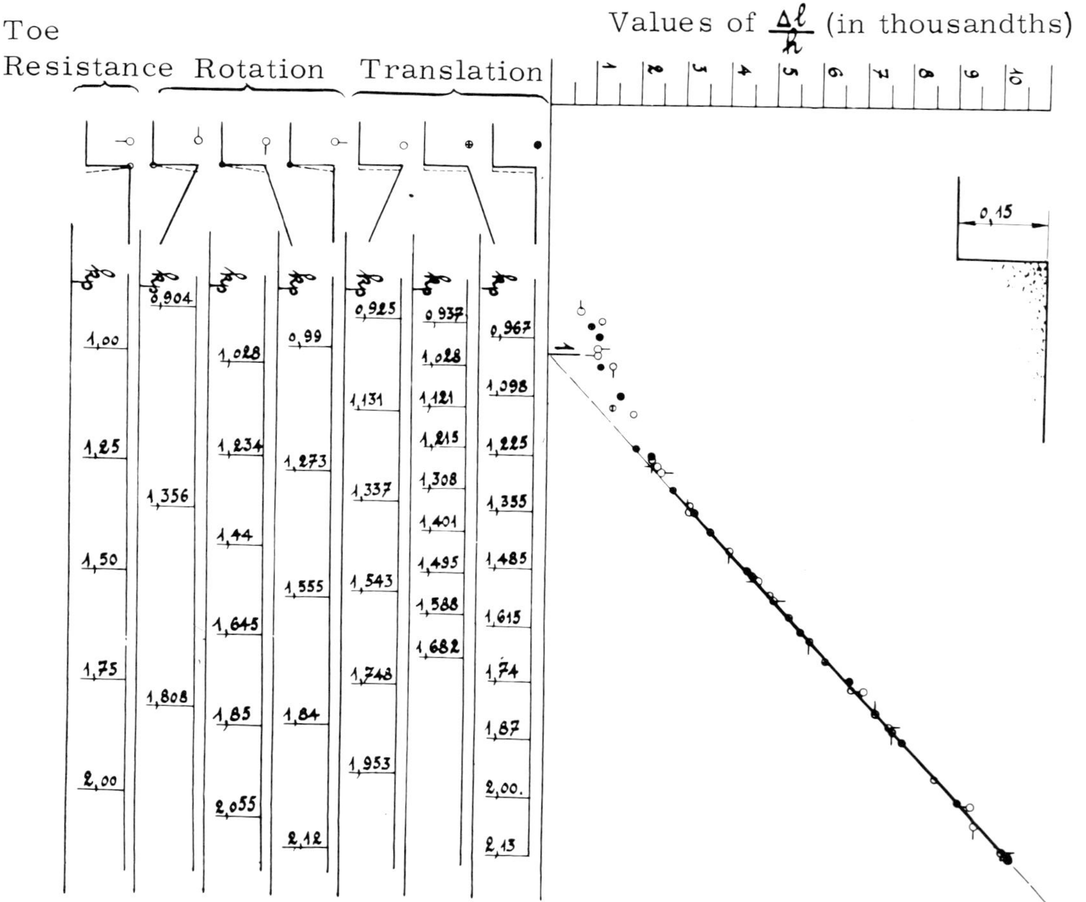

Fig. 59: Translatory and rotational passive resistance and toe resistance.
Deformations as a function of the passive resistance coefficient.
Fine sand.

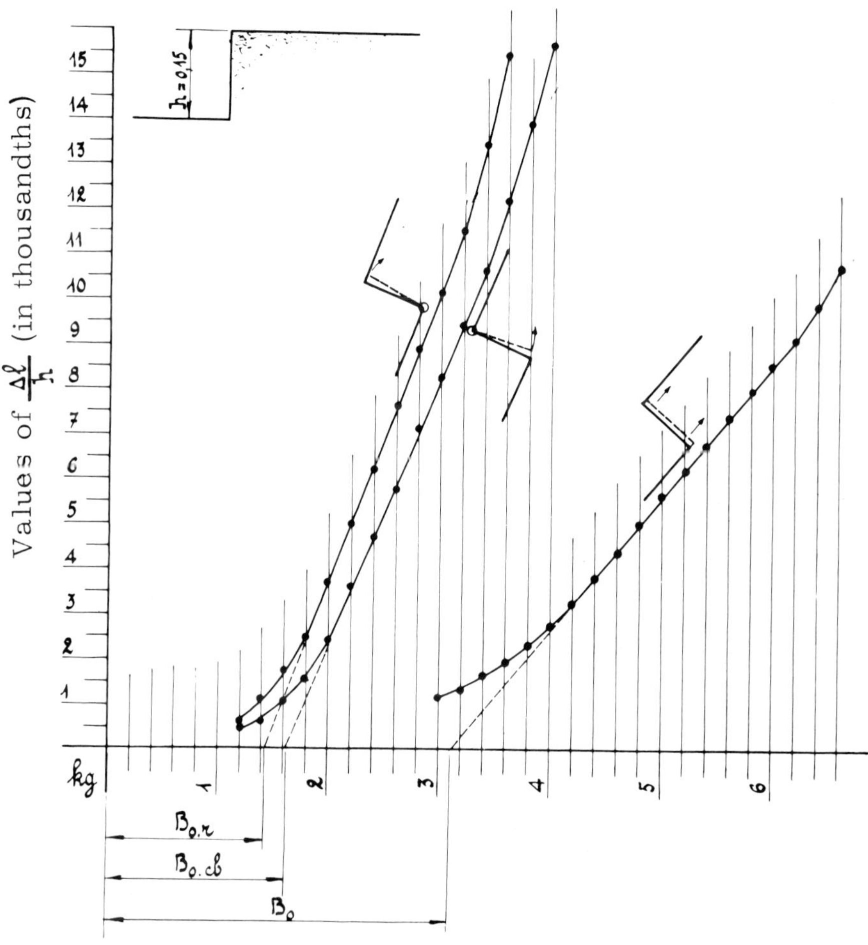

Fig. 60: Translatory and rotational passive resistance and toe resistance.
Deformations as a function of the forces pushing back the diaphragm.
Fine sand.

16.3. Convergence of the Straight Lines Representing the Deformations of a Granular Material Developing Resistance

An inspection of the deformation curves in Figure 26 reveals an important characteristic fact which, when considered in its general implications, proves the deformations associated with granular material developing passive resistance to be due to a definite state of equilibrium of the material.

This equilibrium manifests itself in a constant manner, whatever the shape of the mass of material, with horizontal or with inclined top surface, and whatever form of loading is applied to it, namely, involving translatory or rotational passive resistance or toe resistance.

It appears that the straight lines representing the deformations occurring in the stage of elasto-plastic equilibrium for translatory passive resistance, as plotted in Figure 26, converge at one and the same point M located on the axis of ordinates $\Delta l/h$ (Figure 61).

Now this same phenomenon is observed for the curves in Figure 62 which relate to materials with horizontal or inclined top surface in the case of translatory passive resistance as well as rotational passive resistance and toe resistance.

Furthermore, it emerges that the straight line corresponding to the stage of elasto-plastic equilibrium of material whose top surface is inclined at the lower angle of repose $-\alpha$ and which is loaded under conditions of translatory passive resistance is the same as the line for material whose top surface is inclined at the upper angle of repose $+\alpha$ and which is loaded under conditions of rotational passive resistance.

This observation is indeed in conformity with the fact that, according to the universal formula for the equilibrium of a mass of granular material retained by a diaphragm, the translatory passive resistance coefficient for material whose top surface is inclined at the lower angle of repose is:

$$k_p = \left(\frac{\pi - 2\,\varphi_0}{\pi + 2\,\varphi_0}\right)^2 \cdot \left(\frac{\pi + 2\,\varphi_0}{\pi - 2\,\varphi_0}\right)^2 \cdot \left(1 - \frac{2\,\alpha}{\pi}\right)$$

$$= 1 - \frac{2\,\alpha}{\pi}$$

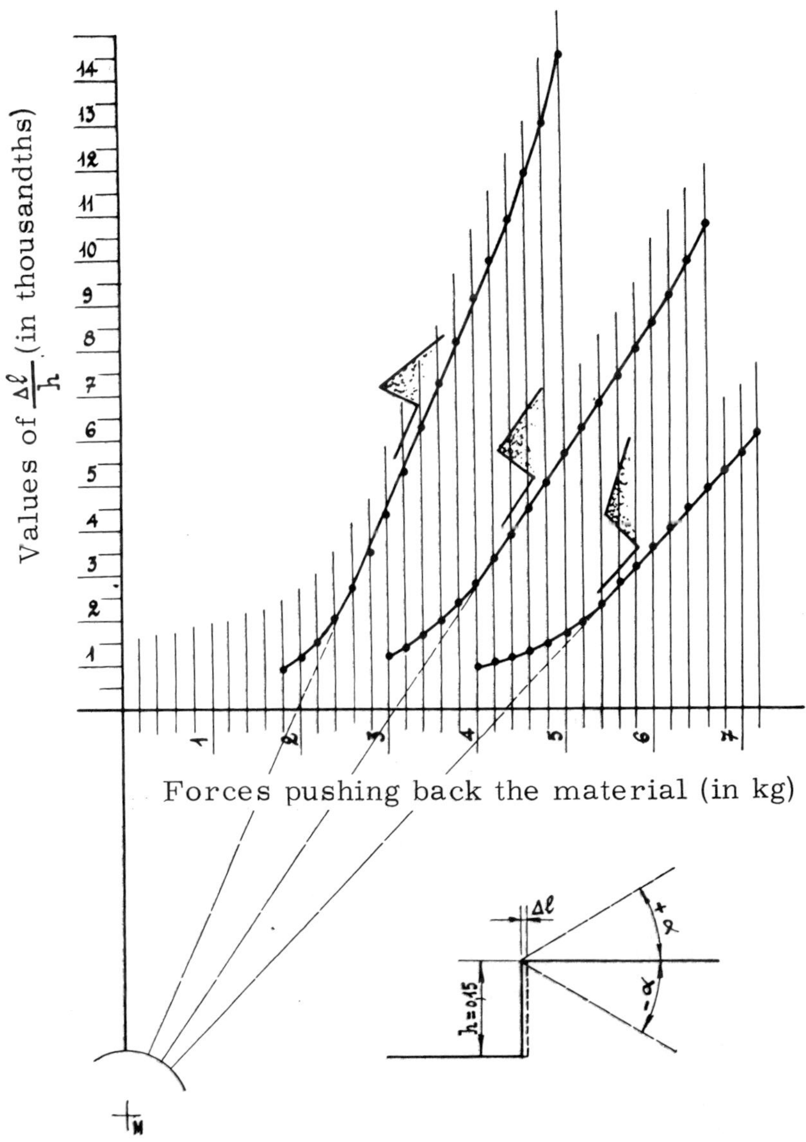

Fig. 61: Stage of elasto-plastic equilibrium. Materials developing translatory passive resistance: Fine sand.

and the rotational passive resistance coefficient for material whose top surface is inclined at the upper angle of repose is:

$$k_p = \left(\frac{\pi - 2\,\varphi_0}{\pi + 2\,\varphi_0}\right)^2 \cdot \frac{\pi + 2\,\varphi_0}{\pi - 2\,\varphi_0} \cdot \left(1 + \frac{2\,\alpha}{\pi}\right)$$

$$= 1 - \frac{2\,\alpha}{\pi} \qquad \text{(bearing in mind that } \varphi_0 \cong \alpha)$$

The passive resistance reaction of the material is therefore the same in both cases:

$$B = \frac{\gamma \cdot h^2}{2} \cdot b \cdot \left(1 - \frac{2\,\alpha}{\pi}\right)$$

and so it stands to reason that the deformations observed in the two sets of experiments are indeed the same.

Important remark

The position of the point M on the vertical axis (axis of ordinates $\Delta l/h$) is especially significant in that the ordinate of that point corresponds accurately to the value of the parameter χ (Section 6.5.).

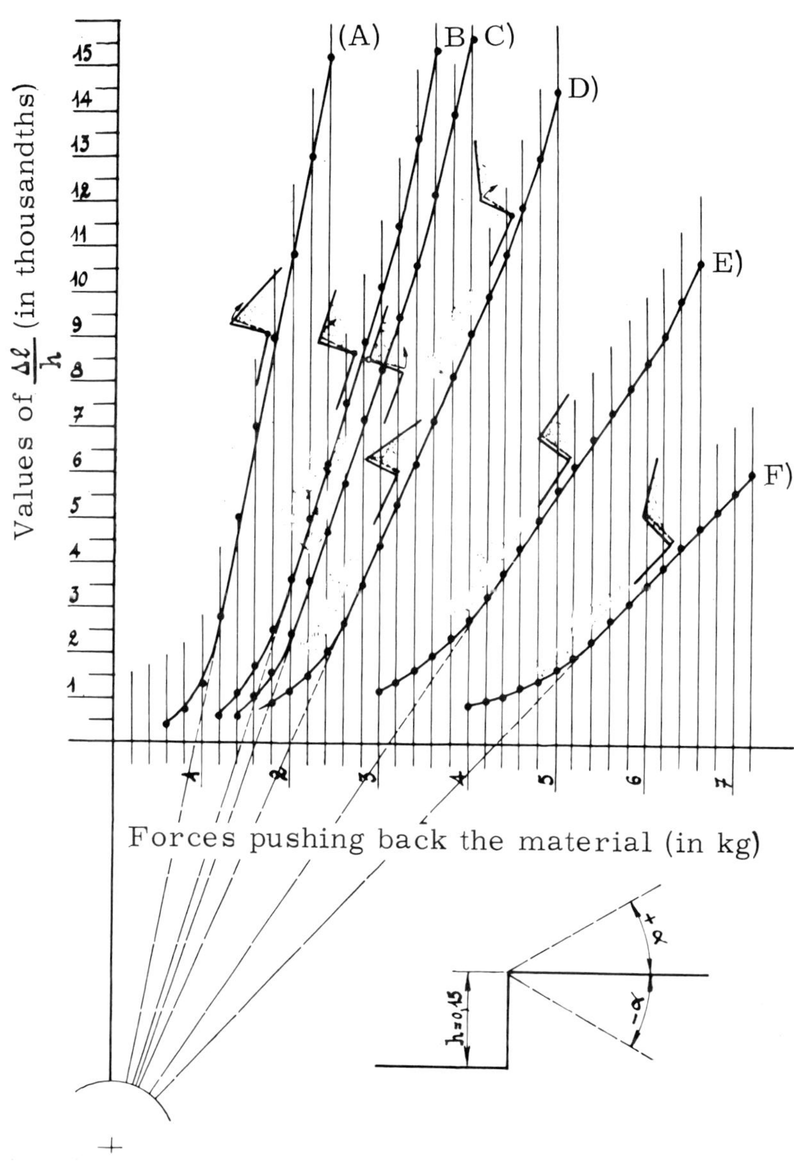

Fig. 62: Stage of elasto-plastic equilibrium. Materials developing translatory or rotational passive resistance or toe resistance: Fine sand.
(A), (B), (D) rotation
(C) toe resistance
(D), (E), (F) translation

17. Summary and Formulas

17.1. Introduction

The following has been established:

1. The condition of the surface of the retaining diaphragm (or in general
   the retaining wall) is of no influence on the value of the horizontal
   component of the passive resistance when the diaphragm is "at rest";
   it is also of no influence on the deformation of the material under the
   action of forces tending to push back the diaphragm against it.

2. No dilatancy of a granular material developing passive resistance manifests itself during the first two stages of equilibrium as defined below.

3. The deformation of a mass of granular material developing passive resistance comprises three characteristic stages with regard to its state of
   equilibrium:

   a. In the first stage the material undergoes adaptation under the effect
      of the as yet relatively small forces tending to push it back upon
      itself.

   b. Then comes the stage of elasto-plastic equilibrium during which the
      deformation increases proportionally to the forces pushing back the
      material; this linearity is a general law of behaviour, irrespective of
      the manner in which the material develops its passive resistance,
      whether translatory or rotational or as toe resistance.

17.2. Stage of Adaptation of the Material

This first stage of behaviour manifests itself when the diaphragm starts to
push back the material it is retaining, i.e., while the forces exerted upon
the latter are still of relatively small magnitude.

The corresponding curve representing the deformations (shown as a thin
line in Figure 63) is enveloped – in the limiting case where the passive

resistance coefficient k_p is less than, or equal to, unity ($k_p \leqslant 1$) – by a straight line (drawn thick in Figure 63) defined by:

$$\frac{\Delta l}{h} = \frac{1\,400}{\gamma} \cdot 3.5\ k_p \qquad \text{(in thousandths)} \qquad (1)$$

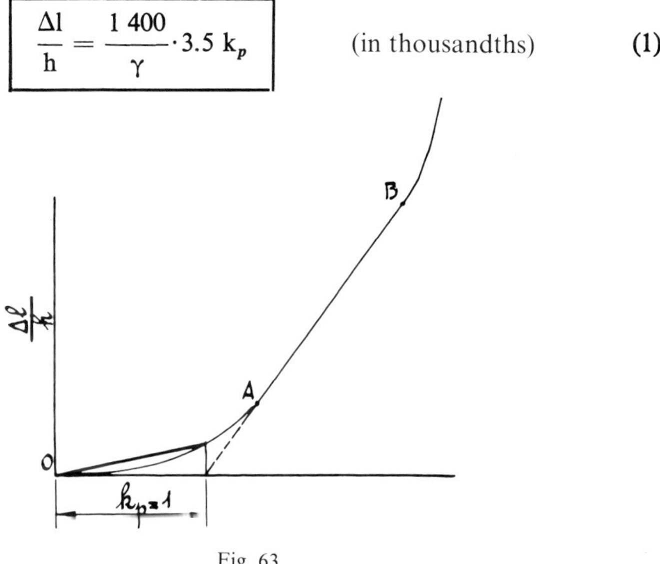

Fig. 63

17.3. Stage of Elasto-Plastic Equilibrium of the Material

During this stage of equilibrium the deformations of the granular material developing translatory or rotational passive resistance or toe resistance conform to the following general law of behaviour:

The deformation of a mass of granular material with horizontal top surface, whether carrying a surcharge or not, loaded under conditions of translatory or of rotational passive resistance or of toe resistance is a linear function of the passive resistance coefficient k_p defined by the ratio between the value of the forces F pushing back the material and the value B_0 of the relative minimum passive resistance, at zero displacement, in the case under consideration.

The straight line representing the deformation (line *AB* in Figure 64) is defined by:

$$\frac{\Delta l}{h} = \chi\,(k_p - 1) \qquad (2)$$

where the value of χ is determined experimentally, while k_p is expressed by:

$$k_p = \frac{F}{B_0} \qquad (3)$$

in which:

$$1 < k_p \leqslant \left(\frac{\pi + 2\,\varphi_0}{\pi - 2\,\varphi_0}\right)^{n'} \qquad (4)$$

$n' = 1$ in the case of loose soil
$n' = 2$ in the case of closely packed or undisturbed soil

and (per linear meter of wall):

$$B_0 = \frac{\gamma \cdot h^2}{2} \cdot \left(\frac{\pi + 2\,\varphi_0}{\pi - 2\,\varphi_0}\right)^{n-2} \cdot \left(1 \pm \frac{2\,\alpha'}{\pi}\right) \cdot \left(1 + \frac{S}{\gamma \cdot h}\right) \qquad (5)$$

where:

$n = 1$ in the case of rotational passive resistance
$n = 2$ in the case of translatory passive resistance
$\alpha' = 0$ in the case of horizontal top surface
$S = 0$ in the case where there is no surcharge

With the aid of formulas (1) to (5) it is possible to solve all problems in connection with determining the deformations affecting any mass of granular material developing passive resistance when the appropriate value of χ for that material has been determined.

If no experimentally determined value, as obtained in a soil testing laboratory, is as yet available for χ, it can permissibly be assumed, e.g., for the purpose of preliminary design calculations, that the straight line representing the deformations in the second stage is expressed (in thousandths) by the following conservative formula (corresponding to the thick line in Figure 64):

$$\frac{\Delta l}{h} = \frac{1\,400}{\gamma} \cdot [3.5 + 10\,(k_p - 1)] \qquad (6)$$

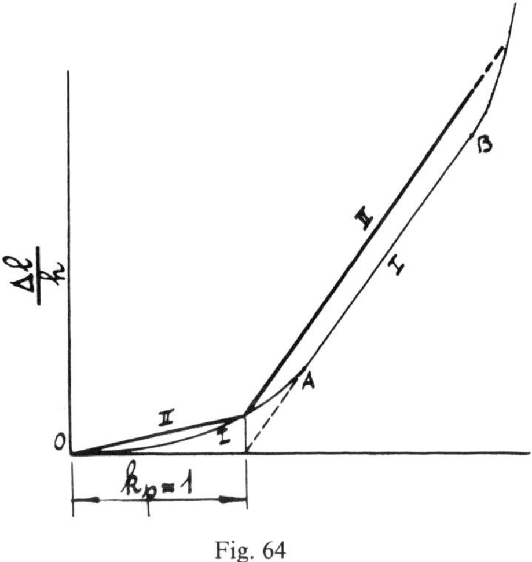

Fig. 64

17.4. Deformations of Vibrated Granular Material or of Closely Packed or Undisturbed Soil

The preceding formulas apply to the deformations of loose granular material developing passive resistance.

These same formulas are applicable to the deformations of vibrated material or to those of closely packed or undisturbed soil, provided that only one-fifth ($\frac{1}{5}$) of the values thus calculated for loose material by means of formulas (1) to (5) are now adopted. Therefore:

$$\frac{\Delta l}{h} = 0.20 \cdot \frac{1\,400}{\gamma} \cdot 3.5 \, k_p \qquad \text{(in thousandths)} \qquad (7)$$

in the case where $k_p \leqslant 1$ or:

$$\frac{\Delta l}{h} = 0.2 \cdot \frac{1\,400}{\gamma} \cdot [3.5 + 10 \, (k_p - 1)] \qquad \text{(in thousandths)} \qquad (8)$$

in the case where:

$$1 < k_p \leqslant \left(\frac{\pi + 2\,\varphi_0}{\pi - 2\,\varphi_0}\right)^2 \qquad (9)$$

17.5. Deformations of Compact Material

In the case of heavily compacted soil, with a reduction in volume of at least 15% as a result of compaction, the deformation to be taken into account is of the following order of magnitude:

$$\boxed{\frac{\Delta l}{h} = \frac{1\,400}{\gamma} \cdot 0.175\, k_p} \quad \text{(in thousandths)} \qquad (10)$$

in which k_p has the value already defined in formula (3).

The deformation value of very compact soil which can carry a compressive stress of 15 bars is of the following order of magnitude:

$$\boxed{\frac{\Delta l}{h} = \frac{1\,400}{\gamma} \cdot 0.02\, k_p} \qquad (11)$$

17.6. Deformations of a Cohesive Mass of Granular Material

In a case where, despite the need for caution that this involves, as has been more particularly discussed in Volume I, the cohesion of soil can permissibly, though as an exception, be taken into account in the design of a temporary structure, the value to be adopted for the thrust F exerted by the soil may be reduced by the following amount (per unit area of surface of the retaining wall):

$$\boxed{(1 - k_a)\, c \cdot \cotg\, \varphi} \qquad (12)$$

where c denotes the cohesion and k_a the thrust coefficient for cohesionless soils.

Under the same conditions the passive resistance will have to be increased by:

$$\boxed{(k_p - 1)\, c \cdot \cotg\, \varphi} \qquad (13)$$

Formulas (11) and (12) are obtained by applying Caquot's theorem of corresponding states (6).

17.7. Stage of Plastic Equilibrium of the Material

In the third stage, namely, that of plastic equilibrium of the material (beyond point B in Figure 16) the deformations increase irregularly as the force tending to push back the material is increased. This behaviour involves a clear presumption of uncertainty for a valid mathematical interpretation and necessitates the adoption of a factor of safety.

18. Calculations for the Design of Sheet Pile Walls and Diaphragm Retaining Walls

18.1. Sheet Piling

In Vol. I, the penetration of sheet piles was determined on the assumption that the active and the passive pressure, or the thrust and the passive resistance, acting on such structures are estimated for the state of equilibrium "at rest", as is indeed laid down in British Code of Practice: *"The pressure to be taken is the value of earth pressure at rest"*.

In the Danish regulations it is specified that if stability calculations for sheet pile walls are based on Rankine's theory, the length of pile penetration obtained by that method should be multiplied by $\sqrt{2}$.

The "at rest" calculation method has the advantage of giving more favourable solutions than those obtained by applying the Danish regulations and guarantees perfect equilibrium of the sheet piling under the best conditions of economy.

However, with the information derived from the experimental work described in the present Vol. II relating to granular materials under passive resistance loading, the deformations of the material which are associated with the displacement of the diaphragm (or, in general, of the retaining wall) can be determined. It is then possible to verify that the solutions obtained for retaining structures designed for the "at rest" condition do indeed satisfy the conditions imposed by the permissible deformations of those structures.

In other words, when the penetration of a sheet pile wall, for example, has been determined by applying the equilibrium formulas for thrust and passive resistance "at rest" so as to ensure perfect stability of the structure, we should then check the value of the deformations associated with the state of elasto-plastic equilibrium of the material developing passive resistance and verify the stresses that are set up in the structural material of the sheet pile wall as a result of such deformations.

This will be further elucidated in the examples presented here.

18.1.1. First Example

We wish to determine the deformations occurring in a sheet pile wall designed for the "at rest" condition, not taking its flexibility into account. It retains a 4 m height of cohesionless granular backfill material exerting the maximum thrust at rest. The density of this material is $\gamma = 1\,700$ kg/m^3; its minimum angle of internal friction is $\varphi_0 = 35°$.

The necessary penetration of the piles is determined by applying the formula given in the chapter entitled "Sheet Piling" in Vol. I:

$$ l = 4.00 \times \cfrac{1}{\cfrac{1}{\mathrm{tg}^{2/3}\left(\cfrac{\pi}{4} - \cfrac{35°}{3}\right)} - 1} $$

$$ = 12.68 \text{ m} \quad \text{say} \quad 12.70 \text{ m} $$

Total length of the sheet piles:

$$ L = 4.00 + 12.70 = 16.70 \text{ m} $$

Thrust exerted by the backfill over the height of the wall:

$$ P = \frac{1\,700 \times \overline{16.70}^2}{2} \times \left(\frac{\pi - 2 \times 35°}{\pi + 2 \times 35°}\right)^2 = 45\,800 \text{ kg} $$

The translatory passive resistance which is developed so as to oppose the horizontal displacement of the toe of the wall ensures the equilibrium of the latter by mobilising a passive resistance coefficient k_{p1} which can be determined by taking equal moments about the toe of the wall:

$$ \frac{45\,800 \times 16.70}{3} \times \frac{1\,700 \times \overline{12.70}^2}{2} \times \frac{12.70}{3} = k_{p1} $$

therefore

$$ k_{p_1} = 0.44 < 1 $$

The corresponding displacement of the toe of the sheet pile wall is (see Sections 17.2., 17.3.):

$$ \Delta l = \frac{12.70}{1\,000} \times \frac{1}{5} \times \frac{1\,400}{1\,700} \times 3.5 \times 0.44 $$

$$ = 0.32 \text{ cm} $$

When this displacement has occurred, the action of the thrust will tend to make the wall lean over by swivelling about its toe; this movement will be resisted by the rotational passive resistance.

The relative minimum rotational passive resistance, for zero displacement, is:

$$B_{0.r} = \frac{1\ 700 \times \overline{12.70^2}}{2} \cdot \frac{\pi - 2 \times 35°}{\pi + 2 \times 35°}$$

$$= 60\ 250\ \text{kg}$$

This passive resistance equilibrates the thrust developed by the backfill and is associated with a passive resistance coefficient k_{p2} which can be determined taking equal moments about the toe:

$$\frac{45\ 800 \times 16.70}{3} = \frac{60\ 250 \times 12.70}{3} \times k_{p2}$$

therefore:

$$k_{p2} = 1$$

This result confirms that the length of penetration previously calculated is indeed in accordance with the state of equilibrium at rest.

The corresponding displacement of the sheet pile wall at the lower ground surface level is:

$$\Delta'l = \frac{12.70}{1\ 000} \cdot \frac{1}{5} \cdot \frac{1\ 400}{1\ 700} \cdot 3.5 \times 1$$

$$= 0.73\ \text{cm}$$

Therefore the top of the wall will undergo a displacement of the following magnitude in relation to its original position:

$$\Delta l + \Delta'l \times \frac{16.70}{12.70}$$

or

$$0.32 + 0.73 \times \frac{16.70}{12.70} = 1.28\ \text{cm}$$

which is less than 1/1 000 of the height of the sheet piling.

Remarks

The foregoing example clearly shows that a sheet pile wall which has been calculated in accordance with the "at rest" equilibrium formulas giving a penetration of the order of three times the retaining height of the wall will undergo only a negligible amount of movement at the top, i.e., the deformation of the sheet piling – or the corresponding unbalanced thrust – is negligibly small.

On the other hand, if the penetration is substantially reduced – say, to only twice the retaining height – the deformation of the sheet piling becomes inadmissibly large, as the second example (below) shows.

Cohesion (if any) may be taken into account in the particular and exceptional case of temporary structures, subject to the conditions explained in Vol. I. In that case the formulas given in Section 17.5. can be applied.

This remark is valid for the various examples presented.

18.1.2. Second Example

We wish to determine the deformations occurring at the top and at the toe of a cantilevered sheet pile wall (i.e., fixed in the soil and not anchored), not taking its flexibility into account. It retains a 4 m height of cohesionless granular backfill material with density $\gamma = 1\,700$ kg/m^3 and minimum angle of internal friction $\varphi_0 = 35°$. The piles forming the wall are given 8.00 m penetration.

Overall height of the sheet pile wall:

$$L = 4.00 + 8.00 = 12.00 \text{ m}$$

Total thrust acting over the height of the wall:

$$P = \frac{1\,700 \times \overline{12.00^2}}{2} \times \left(\frac{\pi - 2 \times 35°}{\pi + 2 \times 35°} \right)^2$$

$$= 23\,700 \text{ kg}$$

The length of penetration necessary – "at rest" – to ensure that the passive resistance developed by the soil over this height (i.e., the penetration) will equilibrate that thrust is well in excess of the 8.00 m penetration actually adopted. This equilibrium with regard to the toe of the wall can be ensured only by the opposing action of the translatory passive resistance character-ised by a coefficient k_{p1} whose magnitude is determined from:

$$\frac{23\ 700 \times 12.00}{3} = \frac{1\ 700 \times \overline{8.00}^2}{2} \times \frac{8.00}{3} \times k_{p1}$$

therefore

$$k_{p_1} = 0.655 < 1$$

Corresponding displacement of the toe of the sheet pile wall (see Sections 17.2., 17.3.):

$$\Delta l = \frac{8.00}{1\ 000} \times \frac{1}{5} \times \frac{1\ 400}{1\ 700} \times 3.5 \times 0.655$$
$$= 3\ cm$$

When this displacement has occurred, the wall finds a new state of equilibrium by swivelling about its toe. This movement of the wall is opposed by the rotational passive resistance.

The relative minimum rotational passive resistance, at zero displacement, is:

$$B_{0.r} = \frac{1\ 700 \times \overline{8.00}^2}{2} \cdot \frac{\pi - 2 \times 35°}{\pi + 2 \times 35°}$$
$$= 24\ 000\ kg$$

This passive resistance can equilibrate the thrust of the backfill only if the passive resistance coefficient k_{p2} has the magnitude determined as follows by taking equal moments about the toe:

$$23\ 700 \times \frac{12.00}{3} = 24\ 000 \times \frac{8.00}{3} \times k_{p_2}$$

therefore

$$k_{p_2} = 1.48$$

The corresponding displacement of the sheet piles at the lower ground surface level is:

$$\Delta'l = \frac{8.00}{1\ 000} \times \frac{1}{5} \times \frac{1\ 400}{1\ 700} \cdot [3.5 + 10\ (1.48 - 1)]$$
$$= 10.95\ cm$$

Thus the top of the wall will, in relation to its initial position, undergo a displacement of the following magnitude:

$$\Delta l + \Delta' l \times \frac{12.00}{8.00}$$

or

$$3 + 10.95 \times \frac{12.00}{8.00} = 19.4 \text{ cm}$$

Remark

This deformation, corresponding to an "unbalanced thrust" developed by the sheet pile wall, is of considerable magnitude and hardly acceptable, even though the penetration is equal to twice the retaining height of the wall. For this reason the use of cantilevered sheet piling, relying simply on its fixity in the ground, is justifiable only in exceptional circumstances and then only if the deformation that occurs is of permissible magnitude. Otherwise the piling should be anchored at the top.

The advantages – in terms of deformations – of using top-anchored sheet pile walls are illustrated by the next two examples.

18.1.3. Third Example

We wish to determine the deformations, and the resulting stress increases, in a retaining wall composed of Larssen V sheet piles anchored at the top and having 8.00 m penetration. The retaining height is 10.00 m and the material retained by the wall is cohesionless undisturbed granular soil with density $\gamma = 1\,700$ kg/m^3 and angle of internal friction $\varphi_0 = 35°$. The tie-rods of the anchorage slope back from the top of the wall at an inclination of 20°.

Preliminary Considerations:

If the sheet pile wall – secured by anchorage at the top – were very rigid, only the toe of the wall would, under the action of the thrust exerted by the soil, undergo an overall displacement. This would be resisted by the reaction of the material in front of the wall, over a height equal to the penetration of the piles. This reaction would correspond to translatory passive resistance. But because of the flexibility of these steel sheet piles it is observed that these – if their length of penetration has been determined by the "at rest" method of calculation – deform elastically by swivelling about the toe of the wall and thus pushing against the material in front of the wall over a height equal to the penetration. This material develops a reaction corresponding to rotational passive resistance.

The calculation procedure adopted for the present example therefore consists in first determining the overall displacement of the toe of the wall in relation to the translatory passive resistance developed over the penetration height and in then considering the base of the wall as fixed and next determining the deformation due to the elasticity of the sheet piling in relation to the rotational passive resistance developed by the material.

Data

The section properties of Larssen V piles per linear metre of sheet pile wall are as follows:

$$\text{Cross-sectional area:} \quad : S \;=\; 300 \text{ cm}^2$$

$$\text{Moment of inertia:} \quad : I \;=\; 50\,900 \text{ cm}^4$$

$$\text{Section modulus:} \quad : \frac{I}{v} = 2\,960 \text{ cm}^3$$

Calculations:

Forces relating to the soil:

Thrust acting over the overall height of the sheet piling:

$$P = \frac{\gamma \cdot h^2}{2} \cdot \left(\frac{\pi - 2\,\varphi_0}{\pi + 2\,\varphi_0} \right)^2$$

$$= \frac{1\,700\,(10.00 + 8.00)^2}{2} \cdot \left(\frac{\pi - 2 \times 35°}{\pi + 2 \times 35°} \right)^2$$

$$= 53\,300 \text{ kg}$$

Translatory passive resistance over the penetration height:

$$B = \frac{\gamma \cdot h^2}{2}$$

$$= \frac{1\,700 \times \overline{8.00}^2}{2}$$

$$= 54\,400 \text{ kg}$$

Since the thrust and the passive resistance are thus seen to be of approximately equal magnitude, the tensile horizontal force to be resisted by the anchorage tie-rods will be determined by the equation for the equilibrium of the moments about the fixed point of attachment of the tie-rods.

Tensile force per linear metre of sheet pile wall:

$$T_0 = \frac{1}{10.00 + 8.00} \cdot \left(53\,300 \times \frac{10.00 + 8.00}{3} - 54\,400 \times \frac{8.00}{3}\right)$$

$$= 9\,725\ \text{kg}$$

Tensile force in the tie-rods (per linear metre of wall):

$$T = \frac{9\,725}{\cos 20°}$$

$$= 10\,350\ \text{kg}$$

Direct (or normal) force in the wall (vertical component of this tensile force):

$$N_1 = 10\,350 \times \sin 20°$$

$$= 3\,540\ \text{kg}$$

Maximum bending moment in the span of the sheet piling, due to the thrust acting on it (triangular load distribution):

$$M_0 = \frac{P \cdot L}{7.8}$$

$$= \frac{53\,300\,(10.00 + 8.00)}{7.8}$$

$$= 123\,000\ \text{kgm}$$

This moment occurs at a distance from the toe of the wall equal to:

$$0.422\,(10.00 + 8.00) = 7.60\ \text{m}$$

The bending moment, at that same distance, due to the passive resistance, whose resultant acts at a distance of 8.00/3 m from the toe, is equal to:

$$M_1 = 54\,000 \times \frac{\dfrac{8.00}{3}\,(10.00 + 8.00 - 7.60)}{10.00 + 8.00}$$

$$\cong 84\,000\ \text{kgm}$$

Resultant bending moment at the section situated at 7.60 m from the toe of the sheet pile wall:

$$123\,000 - 84\,000 = 39\,000 \text{ kgm}$$

Deformations of the Sheet Piles

It will be assumed that the slight elongation of the tie-rods due to the tensile force of 10350 kg per linear metre of sheet pile retaining wall can be neglected.

Displacement of the Toe:

Passive resistance coefficient corresponding to the resistance of 54400 kg developed in opposition to the thrust of 53300 kg:

$$k_p = \frac{53\,300}{54\,400} = 0.98 < 1$$

The displacement at the toe of the wall is therefore:

$$\Delta l = \frac{8.00}{1\,000} \times \frac{1}{5} \times \frac{1\,400}{1\,700} \times 3.5 \times 0.98$$

$$= 0.45 \text{ cm}$$

In terms of the design calculations for the sheet pile wall this is a negligible displacement, being less than $1/1\,000$ of the overall height of the wall.

At the toe the wall can thus, for practical purposes, be regarded as immovably restrained; because of their elasticity, the piles undergo deformation, so that they push against the soil while swivelling about the toe of the wall.

The soil counteracts this deformation of the steel piles by developing a rotational passive resistance whose relative minimum value is:

$$B_{0.r} = \frac{1\,700 \times \overline{8.00}^2}{2} \cdot \frac{\pi - 2 \times 35°}{\pi + 2 \times 35°}$$

$$= 23\,900 \text{ kg}$$

To equilibrate the thrust of 53300 kg is possible only by passive resistance characterised by a coefficient of the following magnitude:

$$k_p = \frac{P}{B_{0.r}} = \frac{53\,300}{23\,900}$$

$$= 2.23 > 1 \quad \text{and} \quad 2.23 < \frac{\pi + 2 \times 35°}{\pi - 2 \times 35°} = 2.27$$

The corresponding rotational deformation is therefore:

$$\Delta l_r = \frac{8.00}{1\,000} \times \frac{1}{5} \times \frac{1\,400}{1\,700} \cdot [3.5 + 10\,(2.23 - 1)]$$

$$= 2.08 \text{ cm}$$

Having regard to the deflection curve of the sheet piles between point A (Figure 65) and point B, i.e., the top and the toe of the wall respectively, it can be assumed with fair approximation that the deflection at C under the action of a fictitious uniformly distributed load p, whose value will be determined, is equal to:

$$y = \frac{1}{EI} \cdot \left[\frac{pL^3 l}{24} + \frac{pl^4}{24} - \frac{pLl^3}{12} \right]$$

hence

$$p = \frac{24\,EIy}{L^3 l + l^4 - 2\,Ll^3}$$

and thus

$$p = \frac{24 \times 2\,100\,000 \times 50\,900 \times 2.08}{18.00^3 \times 8.00 + 8.00^4 - 2 \times 18.00 \times 8.00^3} = 1.65 \text{ kg/cm}$$

Maximum bending moment, corresponding to this load, at distance of 7.60 m from toe:

$$M = \frac{165\,(10.00 + 8.00)^2 \times 0.422 \times 0.578}{2} = 6\,525 \text{ kgm}$$

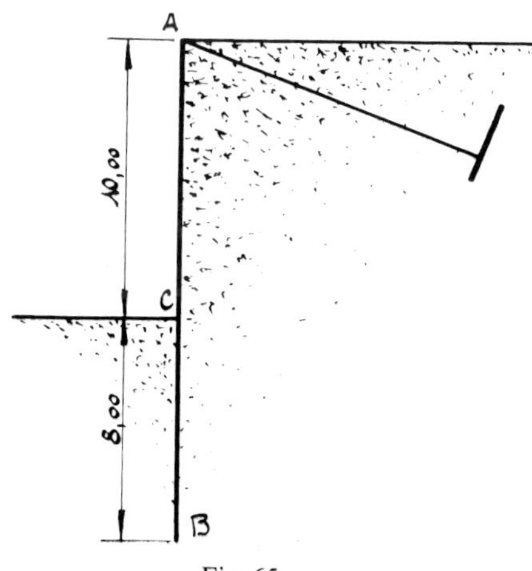

Fig. 65

Stresses in the steel:

Due to the thrust exerted by the soil:

$$\sigma_{a.1} = \frac{39\ 900 \times 100}{2\ 960} = \qquad 1\ 320 \quad \text{bars}$$

Due to the deformation of the piles:

$$\sigma_{a.2} = \frac{6\ 525 \times 100}{2\ 960} = \qquad 221 \quad \text{bars}$$

Due to friction of the soil above the level located at 7.60 m from the toe of the wall:

$$\sigma_{a.3} = \frac{P_2 \cdot \text{tg}\ \theta}{S} \quad (\text{where}\ \theta = \varphi' = \varphi)$$

$$= \frac{1\ 700\ (18.00 - 7.60)^2 \cdot 0.7}{2 \times 300} \cdot \left(\frac{\pi - 2 \times 35°}{\pi + 2 \times 35°} \right)^2 = \quad 41.5\ \text{bars}$$

Due to the vertical component of the pull in the tie-rod:

$$\sigma_{a.4} = \frac{N_1}{S}$$

$$= \frac{3\ 540}{300} = \qquad \underline{11.8\ \text{bars}}$$

$$\text{Total stress in the steel:} \qquad 1\ 594.3 < 1\ 800$$

The sheet pile section chosen for this wall is therefore perfectly suitable, provided that no appreciable corrosion occurs. Otherwise it will be advisable to repeat the foregoing calculation with a heavier pile section to allow for corrosion.

18.1.4. Fourth Example

We wish to determine the deformations and the resulting stresses in a steel sheet pile wall anchored at a distance of 2.00 m below the top and having 6.00 m penetration. The retaining height is 10.00 m and the material retained is cohesionless undisturbed granular soil with density $\gamma = 1\ 700\ \text{kg/m}^3$

and angle of internal friction $\varphi_0 = 35°$. The tie-rods slope back at an angle of 20° (see Figure 66).

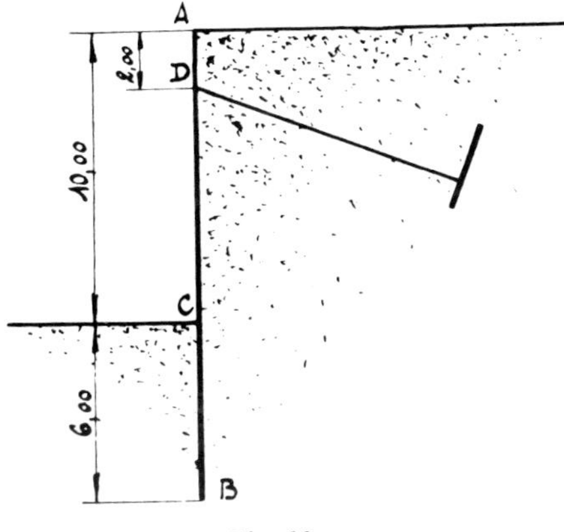

Fig. 66

The section properties of the sheet piles are as given in the preceding example, i.e.:

$$\text{Cross-sectional area:} \quad : S = 300 \text{ cm}^2$$

$$\text{Moment of inertia:} \quad : I = 50\,900 \text{ cm}^4$$

$$\text{Section modulus:} \quad : \frac{I}{v} = 2\,960 \text{ cm}^3$$

Determining the Forces:

The calculation procedure is the same as in the preceding example, replacing L by $L-d$ and taking account of the thrust exerted by the soil above and below the level of the tie-rod attachments to the sheet pile wall.

Unbalanced *thrust* at tie-rod attachment level (D):

$$P_1 = 1\,700 \times 2.00 \left(\frac{\pi - 2 \times 35°}{\pi + 2 \times 35°}\right)^2$$

$$= 660 \text{ kg}$$

Thrust at level of toe of wall (B):

$$P_2 = 1\,700\,(10.00 + 6.00) \left(\frac{\pi - 2 \times 35°}{\pi + 2 \times 35°}\right)^2$$

$$= 5\,280 \text{ kg}$$

Total thrust acting over the height of 14.00 m below tie-rod attachment level:

(a) due to uniformly distributed load:

$$660 \times 14.00 = 9\ 240 \text{ kg}$$

(b) due to triangular load:

$$(5\ 280 - 660) \cdot \frac{14.00}{2} = 32\ 340 \text{ kg}$$

Translatory passive resistance over penetration height:

$$B = \frac{1\ 700 \times \overline{6.00}^2}{2} = 30\ 600 \text{ kg}$$

Tensile Force in the Tie-Rods:

Resultant of the thrust acting above tie-rod attachment:

$$P_3 = \frac{1\ 700 \times \overline{2.00}^2}{2} \cdot \left(\frac{\pi - 2 \times 35°}{\pi + 2 \times 35°} \right)^2$$

$$= 660 \text{ kg}$$

Resultant of the thrust acting below tie-rod attachment:

$$P_4 = \frac{9\ 240}{2} + \frac{32\ 340}{3} - \frac{30\ 600 \times 6.00}{3} \times \frac{1}{8.00 + 6,00}$$

$$= 11\ 100 \text{ kg}$$

Tensile force (pull) in the tie-rods inclined at 20°:

$$F = \frac{660 + 11\ 100}{\cos 20°} \cong 12\ 280 \text{ kg}$$

Vertical component of the pull in the tie-rods (direct or normal force in the sheet piling):

$$N_1 = 12\ 280 \times \sin 20° = 18\ 100 \times 0.342$$

$$\cong 4\ 200 \text{ kg}$$

Maximum bending moment in the span of the sheet piling, due to the thrust acting on it:

Triangular load:

$$M_1 = \frac{32\ 340\ (8.00 + 6.00)}{7.8} = 57\ 900\ \text{kgm}$$

This moment occurs at a distance from the toe equal to:

$$0.422\ (8.00 + 6.00) = 5.90\ \text{m}$$

Uniformly distributed load (moment at 5.90 m from toe B):

$$M_2 = \frac{660 \times 5.90\ (14.00 - 5.90)}{2} = 15\ 800\ \text{kgm}$$

On the other face of the sheet pile wall the reaction (passive resistance) acting over the penetration height produces at the section located 5.90 m from the toe a bending moment M_3 which must be deducted from the moments calculated above:

$$M_3 = 30\ 600 \times \frac{\dfrac{6.00}{3}\ (14.00 - 5.90)}{14.00} = 35\ 300\ \text{kgm}$$

The bending moment acting at 5.90 m from the toe is therefore:

$$M_4 = 57\ 900 + 15\ 800 - 35\ 300 = 38\ 400\ \text{kgm}$$

Soil friction produces a normal force whose magnitude at the section located 5.90 m from the toe will now be calculated:

Thrust at that level:

$$P_5 = \frac{1\ 700\ (10,00 + 6.00 - 5.90)^2}{2} \times \left(\frac{\pi - 2 \times 35°}{\pi + 2 \times 35°}\right)^2 = 16\ 800\ \text{kg}$$

Normal force due to friction with angle $\theta = \varphi$:

$$N_2 = 16\ 800\ \text{tg}\ 35°$$

$$\cong 11\ 800\ \text{kg}$$

Deformation of the Sheet Piles:

Total thrust exerted by the soil over the height extending from the tie-rod attachment D to the toe of the wall:

$$P_5 = \frac{660 + 5\,280}{2} \cdot (8.00 + 6.00) = 41\,580 \text{ kg}$$

Distance from the point of application of the thrust resultant to the toe of the wall:

$$l' = \frac{8.00 + 6.00}{2} \times \frac{2 \times 660 + 5\,280}{660 + 5\,280} = 5.18 \text{ m}$$

Distance from the point of application of the passive resistance resultant to the toe of the wall:

$$l'' = \frac{6.00}{3} = 2.00 \text{ m}$$

Thrust equilibrated by the passive resistance, at the level of the resultant of the latter:

$$P_6 = \frac{41\,580\,(14.00 - 5.18)}{14.00 - 2.00} \cong 30\,500 \text{ kg}$$

Corresponding translatory passive resistance coefficient:

$$k_p = \frac{30\,500 \times 2}{1\,700 \times 6.00^2} \cong 1$$

Displacement of toe of sheet pile wall:

$$\Delta_{l_r} = 6.00 \times \frac{1}{5} \times \frac{1\,400}{1\,700} \times \frac{3.5}{1\,000} \cong 0.35 \text{ cm}$$

In terms of the design calculations for the wall this is a negligible displacement, being less than $1/1\,000$ of the height of the wall from the toe to the point of attachment of the tie-rods.

At the toe the wall can thus, for practical purposes, be regarded as immovably restrained. Because of their elasticity, the piles undergo deformation, so that they push against the soil while swivelling about the toe of the wall.

The soil counteracts this deformation of the steel piles by developing a rotational passive resistance whose minimum value is:

$$B_{0.r} = \frac{1\ 700 \times \overline{6.00}^2}{2} \cdot \frac{\pi - 2 \times 35°}{\pi + 2 \times 35°}$$

$$= 13\ 450 < 30\ 500$$

To equilibrate this thrust of 30 500 kg is possible only by passive resistance characterised by a coefficient of the following magnitude:

$$k_p = \frac{P_6}{B_{0.r}} = \frac{30\ 500}{13\ 450}$$

$$= 2.26 > 1 \quad \text{and} \quad < \frac{\pi + 2 \times 35°}{\pi - 2 \times 35°} = 2.27$$

The corresponding rotational deformation is therefore:

$$\Delta l_r = \frac{6.00}{1\ 000} \times \frac{1}{5} \times \frac{1\ 400}{1\ 700} [3.5 + 10\ (2.24 - 1)]$$

$$= 1.57\ \text{cm}$$

As in the preceding example, it will now be assumed that the deflection at the lower soil surface level (i.e., at C) under the action of a uniformly distributed fictitious load p is:

$$y = \frac{1}{EI} \cdot \left[\frac{p\ (L - d)^3\ l}{24} + \frac{pl^4}{24} - \frac{p\ (L - d)\ l^3}{12} \right]$$

therefore

$$p = \frac{24\ EIy}{(L - d)^3\ l + l^4 - 2\ (L - d)\ l^3}$$

or

$$p = \frac{24 \times 2\ 100\ 000 \times 50\ 900 \times 1.57}{\overline{14.00}^3 \times 6.00 + \overline{6.00}^4 - 2 \times 14.00 \times \overline{6.00}^3} = 3.43\ \text{kg/cm}$$

The corresponding maximum bending moment, at a distance of 5.90 m from the toe of the wall, is:

$$M_5 = \frac{345\ (8.00 + 6.00)^2 \times 0.422 \times 0.578}{2} = 8\ 250\ \text{kgm}$$

Stresses in the Steel:

Due to the thrust exerted by the soil:

$$\sigma_{a.1} = \frac{M_4 \cdot v}{I} = \frac{38\ 400 \times 100}{2\ 960} \cong \qquad\qquad 1\ 290 \text{ bars}$$

Due to the deformation of the piles:

$$\sigma_{a.2} = \frac{M_5 \cdot v}{I} = \frac{8\ 250 \times 100}{2\ 960} = \qquad\qquad 278 \text{ bars}$$

Due to friction of the soil above the level located at 5.90 m from the toe:
Frictional force (with $\theta = \varphi$):

$$F = P \operatorname{tg} \theta = \frac{1\ 700\ (16.00 - 5.90)^2}{2} \cdot \left(\frac{\pi - 2 \times 35°}{\pi + 2 \times 35°}\right)^2 \cdot 0,7$$

$$= 11\ 750 \text{ kg}$$

Resulting stress:

$$\sigma_{a\ 3} = \frac{F}{S} = \frac{11\ 750}{300} \cong \qquad\qquad 39 \text{ bars}$$

Due to the vertical component of the pull in the inclined tie-rod:

$$\sigma_{a.4} = \frac{N_1}{S} = \frac{4\ 200}{300} = \qquad\qquad 14 \text{ bars}$$

Total stress in the steel: $\overline{1\ 621}$ bars

$$1\ 621 < 1\ 800$$

The sheet pile section chosen for this wall is therefore adequate, provided that there is no appreciable corrosion, as in the preceding example.

Remark

The calculations for the sheet pile walls in the second and the third example are based on the assumption that these steel piles are very flexible, so that there is full mobilisation of rotational passive resistance over a height corresponding to the penetration length of the wall.

It may, however, occur that the wall possesses a certain degree of rigidity which limits its deformation. The corresponding value of the passive resistance coefficient k_p will then be lower and therefore the stress in the steel will be smaller than that calculated on the assumption of high flexibility of the sheet piling.

Hence the solutions obtained on the assumption of high flexibility, as in the second and third examples, are on the safe side. If the retaining wall is a structure possessing considerable rigidity, e.g., a reinforced concrete diaphragm wall, the engineer responsible for its design will have to assess in any particular case, and with due regard to this rigidity, whether he can permissibly neglect the flexural deformation of the wall and whether the rotational passive resistance that any such deflection would mobilise is of no practical significance. Otherwise, he should take account of translatory passive resistance in order to determine the displacement of the toe of the wall and to deduce therefrom the resulting increase in the stress acting in the material of the retaining wall.

18.2. Diaphragm Walls

Diaphragm walls (also known as "diaphragm retaining walls" or "perimeter walls") are vertical walls formed in the ground and serving as foundation structures, as water-excluding structures or as ordinary retaining walls.

With regard to the calculation of the deformations that they undergo, and the resulting stresses in the concrete of which they are constructed, diaphragm walls differ from anchored sheet pile walls only in that their moment of inertia is much greater than that of sheet piling and that they are secured by tie-rods or struts constituting a fixed point about which whole wall tends to swivel. In doing so it remains practically plane, i.e., undergoes no deformation, because of its great rigidity.

Thus we here have a case of toe resistance associated with an extremely small amount of displacement of the toe of the wall, even in a case where the angle of internal friction of the soil is relatively small, as will be seen in the fifth example (below) relating to a diaphragm wall used as an ordinary retaining structure.

On the other hand, if the diaphragm wall is used for forming a structural foundation or for carrying superstructures of considerable magnitude, the sixth example will show that the penetration of the wall is determined by the need to reach down into structurally good soil and that the toe will undergo no displacement at all even without mobilisation of passive resistance over the penetration height.

The design calculations will therefore have to be carried out in accordance with the principle envisaged in the fifth example, involving a slight displacement of the toe of the wall, or in accordance with that envisaged in the sixth example, with no displacement of the toe. Which of these alternatives is to be adopted will depend on the loads that the wall has to support.

18.2.1. Fifth Example

Determine the toe displacement of a reinforced concrete diaphragm wall, 0.50 m thick and with an overall height of 14.00 m which comprises 5.00 m penetration length. The wall, which is anchored at the top by means of tie-rods inclined at 15°, retains cohesionless undisturbed alluvial soil having the following mechanical properties: Density $\gamma = 1700$ kg/m³, angle of internal friction $\varphi = 20°$. The penetration has been so determined as to reach good underlying soil with $\varphi = 30°$ (Fig. 67).

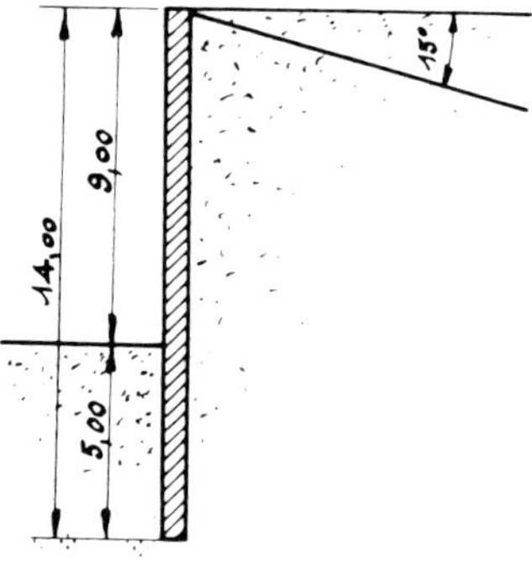

Fig. 67

The forces are calculated per linear metre of wall.

Thrust acting over the height of the wall:

$$P_1 = \frac{1\,700 \times \overline{14.00}^2}{2} \left(\frac{\pi - 2 \times 20°}{\pi + 2 \times 20°} \right)^2 = 67\,500 \text{ kg}$$

Vertical component of soil friction on the wall:

$$N_1 = P_1 \cdot \text{tg } \varphi' = 67\,500 \times 0.364 = 24\,600 \text{ kg}$$

Vertical component of the pull in the tie-rods:

$$N_2 = \frac{P_1}{3} \cdot \text{tg } 15° = \frac{67\ 500}{3} \times 0.268 = 6\ 025 \text{ kg}$$

Dead weight of the wall:

$$N_3 = 14.00 \times 0.50 \times 1.00 \times 2\ 300 = 16\ 100 \text{ kg}$$

Vertical normal (or direct) force acting in the wall:

$$N = 24\ 600 + 6\ 025 + 16\ 100 = 46\ 725 \text{ kg}$$

Frictional force at the underside of the wall, opposing displacement of the toe:

$$F = N \cdot \text{tg } 30° = 46\ 725 \times 0.577 \cong 27\ 000 \text{ kg}$$

Passive resistance developed by the soil, taking the sum of the moments about the point of attachment of the tie-rods:

$$B = \frac{67\ 500 \times \dfrac{14.00 \times 2}{3} - 27\ 000 \times 14.00}{9.00 + \dfrac{5.00 \times 2}{3}} = 20\ 400 \text{ kg}$$

Relative minimum toe resistance, at zero displacement, over the penetration height:

$$B_{0.cb} = \frac{1\ 700 \times \overline{5.00}^2}{2} \cdot \frac{\pi - 2 \times 20°}{\pi + 2 \times 20°} = 13\ 550 \text{ kg}$$

Passive resistance coefficient to be taken into account for equilibrating the passive resistance of 20 400 kg:

$$13\ 550 \times k_p = 20\ 400$$

and therefore

$$k_p = \frac{20\ 400}{13\ 550} = 1.51 > 1$$

Displacement of the toe of the wall:

$$\Delta l = \frac{5.00}{1\ 000} \cdot \frac{1}{5} \cdot \frac{1\ 400}{1\ 700} \cdot [3.5 + 10\ (1.51 - 1)] = 0.71 \text{ cm}$$

Since the rotation involving the need to take account of toe resistance occurs about the point of tie-rod attachment situated far above the upper level of action of the passive resistance over the penetration height of the wall, it follows that the actual phenomenon must be intermediate between the translatory passive resistance and the toe resistance.

If, for the sake of simplifying the calculations, we take the actual passive resistance as equal to the translatory passive resistance, the corresponding coefficient k_p would conform to:

$$20\ 400 = \frac{1\ 700 \times \overline{5.00}^2}{2} \cdot k_p$$

and therefore

$$k_p = 0.96 < 1$$

The displacement of the toe of the wall would thus be:

$$\Delta l = \frac{5.00}{1\ 000} \times \frac{1}{5} \times \frac{1\ 400}{1\ 700} \times 3.5 = 0.29 \text{ cm}$$

It is seen that in both cases the displacement of the toe of the wall is insignificant because of the considerable friction developed on the underside, i.e., on the bearing surface of the wall on the soil underneath.

In cases where diaphragm walls serve as an integral feature of foundations, generally for large structures, they are subjected to very heavy vertical loads. From practical experience it is known that the design of such walls is much simplified, if the stability of the toe is normally ensured by the friction between the base of the wall and the bearing surface on the underlying soil. In such circumstances no displacement of the toe will occur.

It is therefore unnecessary to consider the passive resistance developed over the penetration height of the wall as was done in the sheet piling calculations. Instead, the penetration is calculated as the end span of a continuous slab which bends between its two supports, these being, respectively, the extreme bottom end of the wall (where it bears on the soil) and the first row of struts or lower anchorage level.

The sixth example, given below, will show that the loads acting on a diaphragm wall, provided that they are of substantial magnitude, are sufficient to ensure the stability of the toe. The actual penetration to be adopted will depend only on the need to extend down to good loadbearing soil.

18.2.2. Sixth Example

Check that the toe undergoes no displacement in the case of 0.50 m thick reinforced concrete diaphragm wall, 14.00 m overall height, 2.00 m penetration, strengthened by reinforced concrete floors serving to strut it (see Fig. 68). The penetration is sufficient to reach loadbearing soil consisting of a limestone bed which can safely be loaded to a bearing pressure of 20 bars and has an angle of internal friction of 38°. The diaphragm wall supports a superstructure whose bearing reaction, including the loads from the floors, is 30 tons per linear metre of wall. The mechanical properties of the soil are the same as in the preceding example.

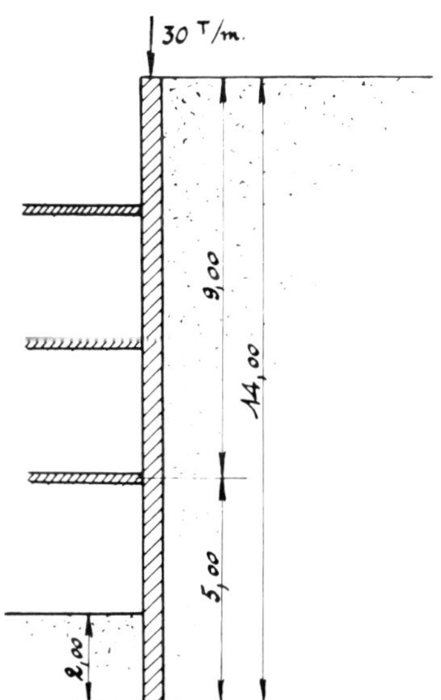

Fig. 68

The forces are calculated per linear metre of wall.

Thrust acting over the height of the wall:

$$P_1 = \frac{1\,700 \times 14.00^2}{2} \cdot \left(\frac{\pi - 2 \times 20°}{\pi + 2 \times 20°}\right)^2$$

$$= 67\,500 \text{ kg}$$

Thrust acting on the front of the wall, over the penetration height:

$$P_2 = \frac{1\,700 \times \overline{2.00}^2}{2} \cdot \left(\frac{\pi - 2 \times 20°}{\pi + 2 \times 20°}\right)^2$$

$$= 1\,380 \text{ kg}$$

Vertical component of soil friction on the back and on the front of the wall:

$$N_1 = (P_1 + P_2) \, \mathrm{tg} \, 20°$$
$$= (67\,500 + 1\,380) \cdot 0.364$$
$$\cong 25\,070 \text{ kg}$$

Dead weight of the wall:

$$N_2 = 14.00 \times 0.50 \times 1.00 \times 2\,300$$
$$= 16\,100 \text{ kg}$$

Loads due to construction and to the superstructure:

$$N_3 = 30\,000 \text{ kg}$$

Vertical normal (or direct) force at the toe of the wall:

$$N = N_1 + N_2 + N_3$$
$$= 25\,070 + 16\,100 + 30\,000$$
$$= 71\,170 \text{ kg}$$

Pressure on the foundation soil:

$$p = \frac{71\,170}{50 \times 100}$$
$$= 14.25 \text{ bars} < 20$$

Frictional force at the underside of the wall, opposing displacement of the toe:

$$F = N \cdot \mathrm{tg} \, 38°$$
$$= 71\,170 \times 0.781$$
$$\cong 55\,500 \text{ kg}$$

Thrust developed by the soil at the lower floor or strutting level:

$$p_1 = 1\,700 \times 9.00 \left(\frac{\pi - 2 \times 20°}{\pi + 2 \times 20°} \right)^2$$
$$= 6\,200 \text{ kg}$$

Thrust developed by the soil at toe level:

$$p_2 = 1\,700 \times 14.00 \left(\frac{\pi - 2 \times 20°}{\pi + 2 \times 20°} \right)^2$$

$$= 9\,650 \text{ kg}$$

Total thrust on the wall below the lower floor or strutting level:

$$P_3 = \frac{6\,200 + 9\,650}{2} \times 5.00 = 39\,625 \text{ kg}$$

Distance from the point of application of the resultant of the thrust P_3 to the lower floor or strutting level:

$$d = 5.00 - \frac{5.00}{3} \times \frac{2 \times 6\,200 + 9\,650}{6\,200 + 9\,650} = 2.68 \text{ m}$$

Moments due to the thrusts P_2 and P_3 about the lower floor or strutting level:

$$M_1 = 39\,625 \times 2.68 - 1\,380 \left(3.00 + \frac{2,00 \times 2}{3} \right)$$

$$= 100\,320 \text{ kgm}$$

Stabilising moment corresponding to the friction developed by the loaded wall:

$$M_2 = 55\,500 \times 5.00$$

$$= 277\,500 \text{ kgm} > 100\,320$$

i.e., the factor of safety is $\quad : \dfrac{277\,500}{100\,320} \cong 2.77$

The wall is therefore stable on its bearing surface, and its toe will undergo no displacement. This result is obtained without taking account of the passive resistance developed by the soil over the penetration height.

Remark

The foregoing calculation applies to the completed wall in actual use, i.e., performing the structural and loadbearing function for which it is intended. The following calculation shows, however, that the conclusion as to the stability of the toe remains valid even when the wall is not yet subjected to load from the superstructure.

Thus, in the case where the normal (or direct) force in the wall is due only to its dead weight to the vertical component of the friction of the soil in contact with it, we have:

$$N_1 + N_2 = 25\,070 + 16\,100$$
$$= 41\,170 \text{ kg}$$

therefore the frictional force at the underside of the wall, opposing displacement of the toe, is:

$$F = (N_1 + N_2) \text{ tg } 38°$$
$$= 41\,170 \times 0.781$$
$$\cong 32\,200 \text{ kg}$$

hence the stabilising moment is:

$$M_2 = 32\,200 \times 5.00 = 161\,000 > 100\,320$$

The stability of the toe of the wall is therefore ensured even in these circumstances, the factor of safety being:

$$\frac{161\,000}{100\,320} \cong 1.61$$

19. Program for Design by Electronic Computer

The foregoing examples show that the complete analysis of a sheet pile wall or a diaphragm wall is relatively laborious.

It is even more so if the soil is stratified, saturated and subjected to a surcharge, and the efforts in seeking the most economical design then become particularly complex.

Accordingly, it was felt to be essential to develop a computer program that could suitably be used for solving any problem associated with sheet pile and diaphragm wall design. For this reason we have incorporated the various formulas of the present book into a program for use by contractors and consulting engineers concerned with such design work.

This program comprises the computation of thrust/passive resistance curves, bending moments and shear forces as affecting a retaining wall, a sheet pile wall and a diaphragm wall, based on the "at rest" method of design set forth in this book, for maximum thrust, and taking account of the deformations connected with passive resistance limited, to be on the safe side, to the elasto-plastic stage of equilibrium of the material considered.

The method is applicable to granular materials:
- whether cohesive or not
- with horizontal or inclined top surface
- with or without uniformly distributed surcharge
- with or without local surcharge
- whether saturated with water or not
- loose or compact
- homogeneous or stratified

20. Determining the Minimum Angle of Internal Friction of a Granular Material

In Volume I we interpreted the results of our tests on scale models and embodied them in simple formulas as functions of the value of angle of repose α taken as being equal to the minimum angle of internal friction φ_0 of the test material.

Thus the value of the total thrust P exerted on a vertical diaphragm (or in general: a retaining wall) of height h by a granular material whose bulk density is γ and angle of repose is α is expressed by the formula:

$$P = \frac{\gamma \cdot h^2}{2} \cdot \left(\frac{\pi - 2\,\alpha}{\pi + 2\,\alpha} \right)^2$$

or alternatively by the equivalent formula:

$$P = \frac{\gamma \cdot h^2}{2} \cdot \left(\frac{\pi - 2\,\varphi_0}{\pi + 2\,\varphi_0} \right)^2$$

We know that the assumption that the angle of repose is equal to the minimum angle of internal friction is considered to be merely an approximation by soil mechanics engineers; but we must emphasize that this is a favourable approximation for determining the maximum values of the thrust, as these will thus be estimated on the safe side, as will be seen in Section 20.6.

We would like to recall that already in 1942 Buisson (7) indicated the value of the minimum angle of internal friction of a granular material to be very close to the angle of repose thereof, and in 1948, Terzaghi and Peck (8) pointed out that:

"The maximum value of the thrust and the minimum value of the passive resistance are attained when the angle of internal friction φ of the granular material is minimum; it then corresponds to the angle of repose which is characteristic of the limit equilibrium of the material in itself."

We wished to obtain fresh confirmation of the validity of this by means of the experiments described below.

20.1. Determination of the Minimum Angle of Internal Friction φ_0 of a Granular Material

The appliances most widely employed for determining the shearing strength of a granular material as a function of the internal friction thereof are the CASAGRANDE shear box and the triaxial compression apparatus.

Unfortunately, for comparing the value of the minimum angle of internal friction φ_0 with that of the angle of repose α these two appliances are affected by errors which are inherent in their design.

In the case of the shear box the shearing force applied to the cover thereof and the opposing force developed by the bottom half are both transmitted through the walls of the box. As a result, a couple is developed which tends to lift the cover and to increase the shearing strength, which accounts for certain irregularities detected in the results obtained.

On the subject of the shear box test Mayer (9) has the following to say:

"The causes of indeterminacy oblige us to have, for each sample, a sufficient number of tests to enable any irregularities to be detected and nevertheless to have enough accurate points to draw conclusions from the measurements."

As for the triaxial testing apparatus, Caquot and Kérisel (10) have called attention to the snag – a major one, in our opinion – presented by the influence of the membrane that encloses the specimen and which "though very thin, has an influence on the experimental results by imparting rigidity to it."

This goes to show the difficulties hitherto encountered by investigators in determining, for the limit condition, the minimum angle of internal friction of any particular granular material. In order to overcome these various drawbacks we devised a procedure that enabled the material to be relieved of all stress liable to disturb its natural equilibrium.

The device used for the purpose is as follows:

A mass of granular material B with horizontal top surface is placed on a fixed horizontal table A (Fig. 69). In order to measure the minimum angle of internal friction of this material we place on the surface thereof a rough plate C to which loads Q are applied. Also, a horizontal force is applied to this plate C and is increased until the latter slides on the material.

Let F be the force which initiates the sliding of the plate; the coefficient of friction that we wish to determine is then:

$$\operatorname{tg} \varphi = \frac{F}{Q}$$

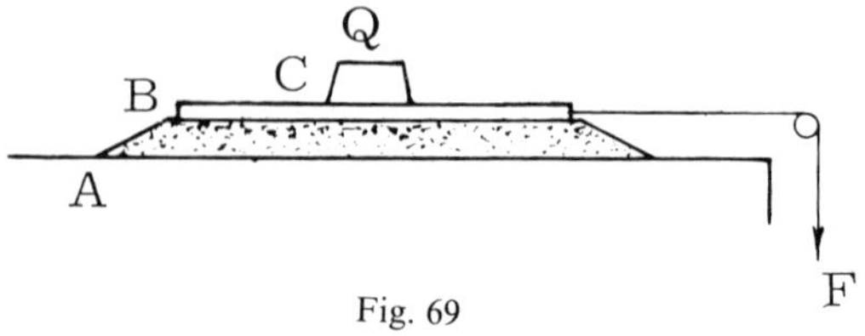

Fig. 69

20.2. Experiments

The face of the plate C in contact with the granular material whose minimum angle of internal friction we wish to determine is roughened beforehand by being provided with a glued-on coating of the test material, so that at the instant when the plate slides under the action of the force F there is grain-to-grain friction of this granular material.

Weights of $Q = 2.5, 5, 10$ and 15 kg are placed on the plate C, and the tensile force F acting horizontally on it is gradually increased until sliding occurs.

The displacements of the plate under the effect of these gradually increasing values of the horizontal force are measured with the aid of a dial gauge with an accuracy of 1 micron.

In a first series of tests the tensile force was exerted by the successive application of loads equal to $1/10, 2/10 \ldots n/10$ of the vertical load applied to the plate C.

The test results were satisfactory and showed less than 2% scatter, which is less than that ordinarily obtained in conventional shear tests. The curves representing these results are regular from the commencement of movement of the plate until it slides.

However, tests of this kind, as reported in the following (loads $Q = 2.5$ kg and $Q = 5$ kg), are relatively time-consuming to perform because it takes time for the material to become stabilised after each increment of the force F is applied. For this reason a second series of tests was performed, but now the force F was increased smoothly and continuously from zero to the value at which sliding of the plate occurs. It was found that this latter value was precisely the same as that obtained in the first series of tests, i.e., with successive incremental increases of F.

These „continuous" tests have the advantage if being much quicker to perform and, since they give the same results as the "discontinuous" ones, they were the preferred procedure for further experimental research. The tensile force applied to the plate C was produced by pouring water into a container suspended from the pull wire. The water was fed at a constant rate by means of a siphoning system.

The properties of the granular materials employed in the tests will be recapitulated here (see Section 3):

Sands:
Fine sand from Garoupe-Antibes:

$$\gamma = 1\ 380 \text{ kg/m}^3 \qquad \alpha = 33° \ 40'$$

Crushed stone sand from Var:

$$\gamma = 1\ 420 \text{ kg/m}^3 \qquad \alpha = 36° \ 30'$$

Sand from the Seine:

$$\gamma = 1\ 550 \text{ kg/m}^3 \qquad \alpha = 34° \ 40'$$

Cereals:
Millet:

$$\gamma = \ \ \ 730 \text{ kg/m}^3 \qquad \alpha = 28° \ 30'$$

Denatured hard wheat:

$$\gamma = \ \ \ 790 \text{ kg/m}^3 \qquad \alpha = 31° \ 30'$$

The tests revealed the following facts:

1. In the case of the sands there occurs an initial and very distinct sliding of the plate C, followed at once by stabilisation; then, with increasing tensile force, comes a second sliding movement, which continues thereafter even without further increase of the force.

   Comparison of the test results reported below with the results of conventional tests suggests that the first sliding movement corresponds to the minimum friction and that the second corresponds to the maximum friction.

2. In the case of the cereals the initial sliding movement is not followed by any stabilisation and it continues without further increase of the force.

20.2.1. First Test Series

The following tables give the values of the horizontal displacements Δl when the plate C, carrying a vertical load Q, is subjected to horizontal tensile forces of magnitude $Q/10, 2Q/10 \ldots nQ/10$.

From the test results given in Table 30 and Figure 70 it can be deduced that the minimum coefficient of friction is:

$$\text{tg } \varphi_0 = \frac{1.688}{2.500} \cong 0.675 \quad \text{and therefore:} \quad \varphi_0 = 34° \, 02' > \alpha$$

and that the maximum coefficient of friction is:

$$\text{tg } \varphi_{max} = \frac{1.926}{2.500} = 0.770 \quad \text{and therefore:} \quad \varphi_{max} = 37° \, 35'$$

Applied Forces F (in kg)	1st Test	2nd Test	3rd Test	Averages
0.250	0.3	0.4	0.4	0.4
0.500	5.5	5.6	5.6	5.6
0.750	22.0	23.5	22.4	22.6
1.000	48.0	51.6	50.5	50.0
1.250	89.0	93.0	91.2	91.0
1.500	147.0	162.0	153.3	154.1
Force at First Sliding	1.670	1.705	1.690	1.688
	$(\Delta l = 198)$	$(\Delta l = 221)$	$(\Delta l = 212)$	$(\Delta l = 210)$
Force at Failure (in kg)	1.840	1.930	2.010	1.926

Table 30

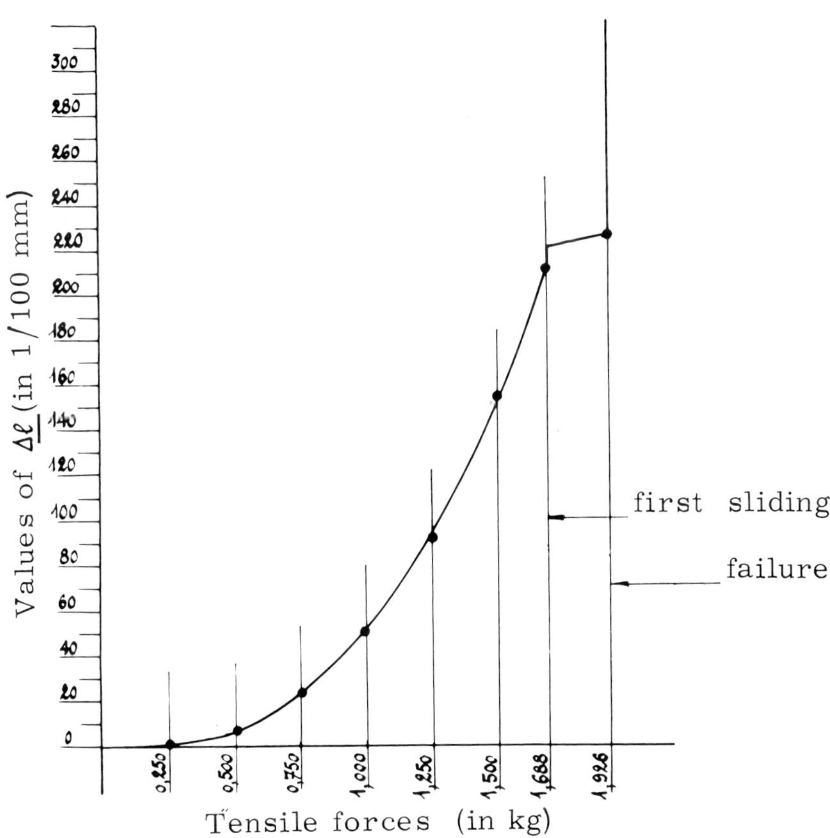

Fig. 70: Material: Fine sand from Garoupe; $Q = 2.5$ kg.

From the test results given in Table 31 and Figure 71 it can be deduced that the minimum coefficient of friction is:

$$\text{tg } \varphi_0 = \frac{3.413}{5.000} \cong 0.683 \quad \text{and therefore:} \quad \varphi_0 = 34° \, 20' > \alpha$$

and that the maximum coefficient of friction is:

$$\text{tg } \varphi_{max} = \frac{4.010}{5.000} = 0.802 \quad \text{and therefore:} \quad \varphi_{max} = 38° \, 45'$$

For loads $Q = 10$ kg and $Q = 15$ kg the curves representing the displacements are similar in character to those obtained in the preceding tests.

The minimum coefficient of friction is very close to that obtained with loads $Q = 2.5$ kg and $Q = 5$ kg, but the maximum coefficient of friction is rather substantially different.

Applied Forces F (in kg)	Valeurs de Δl en 1/100 de mm			
	1st Test	2nd Test	3rd Test	Averages
0.500	0	0	0	0
1.000	1.8	2.0	2.0	1.9
1.500	14.7	14.4	14.8	14.6
2.000	41.0	38.0	39.6	39.5
2.500	78.4	76.0	77.8	77.4
3.000	133.8	132.9	133.3	133.3
Force at First Sliding	3.410 ($\Delta l = 225$)	3.425 ($\Delta l = 218$)	3.405 ($\Delta l = 212$)	3.413 ($\Delta l = 218$)
Force at Failure (in kg)	4.080	3.985	3.965	4.010

Table 31: Material: Fine sand from Garoupe; $Q = 5$ kg

For a load $Q=10$ kg the first sliding movement, in three consecutive tests, was observed to occur at the following values of the tensile force F: 6.975; 6.915 and 6.960 kg, i.e., an average value of 6.950 kg, corresponding to:

$$\text{tg } \varphi_0 = \frac{6.950}{10.000} = 0.695 \text{ and therefore: } \varphi_0 = 34° \; 50'$$

Similarly, for $Q=15$ kg the forces F corresponding to the first sliding movement were: 10.485, 10.470 and 10.510 kg, i.e., an average value of 10.486 kg, corresponding to:

$$\text{tg } \varphi_0 = \frac{10.486}{15.000} \cong 0.700 \text{ and therefore: } \varphi_0 = 35°$$

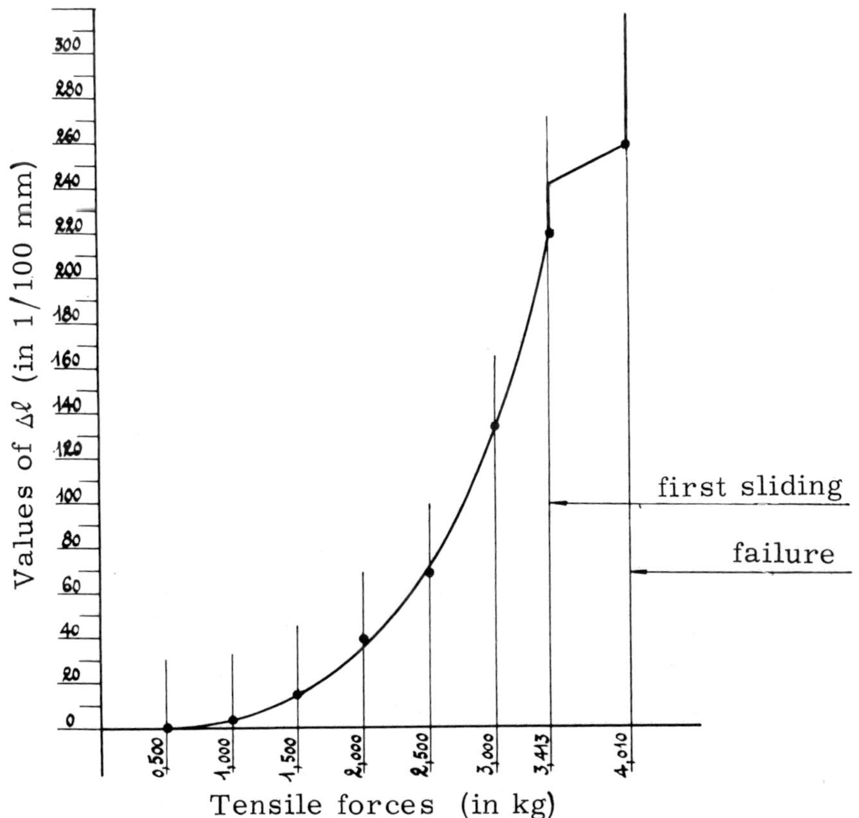

Fig. 71: Material: Fine sand from Garoupe; $Q=5$ kg.

As for the results of the tests to failure, when the plate C is pulled along continuously without further increase in the magnitude of the tensile force acting on it, these show a degree of scatter from which it can only be inferred that the maximum coefficient of friction is not so definite as in the two preceding cases, but that it is nevertheless larger than the coefficients of friction corresponding to $Q=2.5$ kg and $Q=5$ kg.

In view of the remark made in Section 20.2., no tests in the second series were performed with cereals.

20.2.2. Second Test Series

In this second series the tensile force is applied smoothly and continuously, not in successive increments, as has been explained in Section 20.2.

The displacements Δl can be accurately observed by taking dial gauge readings at regular intervals, e. g., every minute. The magnitude of the load applied – water poured into a container at a regular rate of flow – is known at all times.

The results of these tests are given below. The figures of one test for each material are reported.

Note

The rate of discharge of the water into the container is constant during the course of any one and the same test, but may vary from one test to another, depending on the diameter of the inlet tube adopted for a particular test.

Test on Seine Sand with a Load of $Q=5$ kg:

An initial load of $Q=0.600$ kg, due to the weight of the container, is applied and produces a first displacement of the plate C. Then water is allowed to flow into the container, and at one-minute intervals the dial gauge indicating the displacements Δl of the plate is read (Table 32 and Fig. 72).

The load which caused first sliding movement was
3.470 kg
to which corresponds the minimum coefficient of friction:

$$\text{tg } \varphi_0 = \frac{3.470}{5.000} = 0.694 \quad \text{and therefore:} \quad \varphi_0 = 34° \ 45'$$

Time (in min)	$\triangle$ l (in 1/100 mm)	Time (in min)	$\triangle$ l (in 1/100 mm)
		18	68.7
1	0.7	19	76.7
2	1.2	20	85.0
3	2.8	21	95.0
4	4.8	22	106.5
5	7.2	23	118.1
6	10.7	24	130.6
7	13.1	25	145,4
8	16.2	26	164.6
9	19.8	27	184.0
10	23.6	28	202
11	27.6	29	227
12	31.7	30	250
13	37.3	31	278
14	42.8	32	327
15	50.2	33	335
16	54.6	34	344
17	61.0	35	347
		36	350
		38′ 30″	failure

Table 32: Material: Seine sand; $Q = 5$ kg

The load which caused failure:
$$4.170 \text{ kg}$$
to which corresponds the maximum coefficient of friction:

$$\text{tg } \varphi_{max} = \frac{4.170}{5.000} = 0.834 \quad \text{and therefore:} \quad \varphi_{max} = 40° \, 20'$$

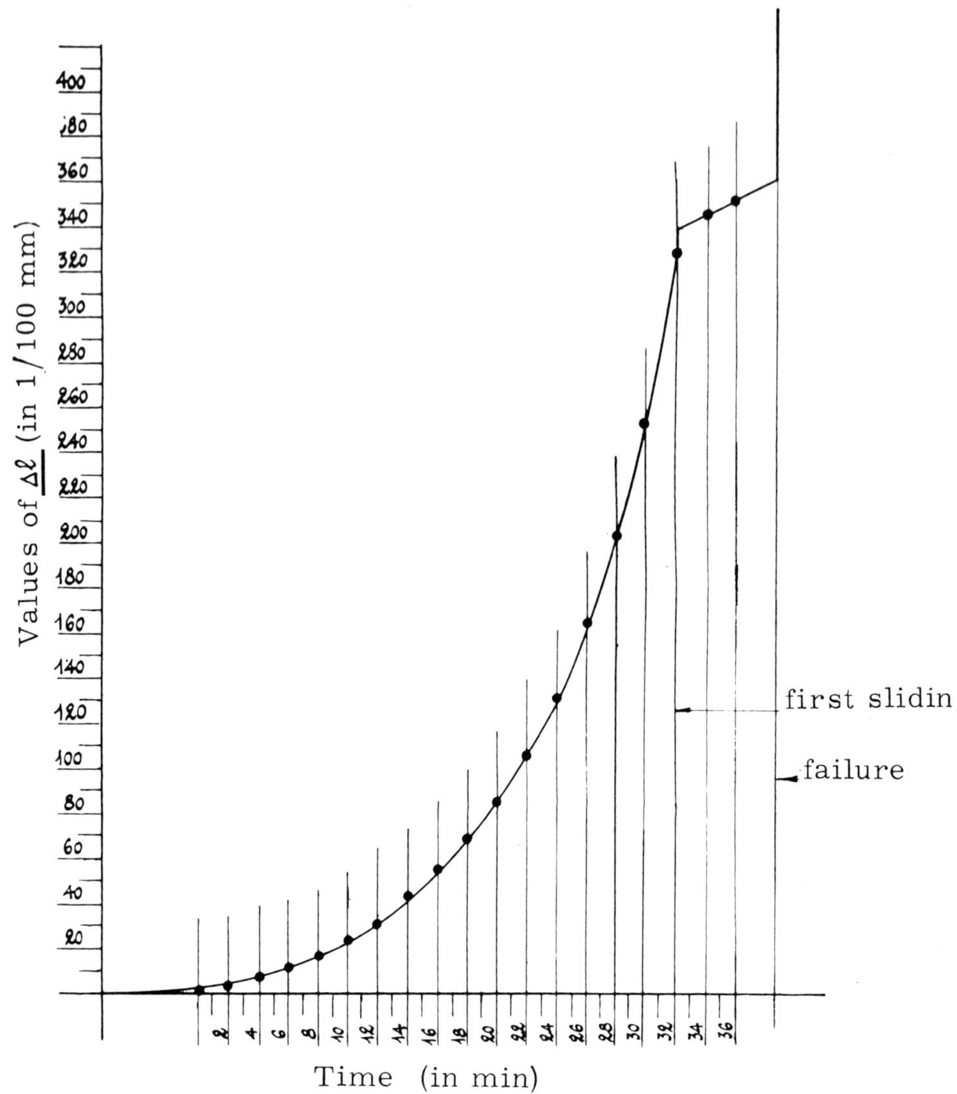

Fig. 72: Material: Seine sand; $Q = 5$ kg.

Test on Var crushed stone sand with a load of $Q=5$ kg:
The test conditions were the same as in the preceding case, the initial load
again $Q=0.600$ kg (Table 33 and Fig. 73).

The load that caused first sliding movement:
$$3.830 \text{ kg}$$
to which corresponds the minimum coefficient of friction:

$$\text{tg } \varphi_0 = \frac{3.430}{5.000} = 0.686 \text{ and therefore: } \varphi_0 = 34° \ 25'$$

The load that caused failure:
4.110 kg
to which corresponds the maximum coefficient of friction:

$$\text{tg } \varphi_{max} = \frac{4.110}{5.000} = 0.822 \text{ and therefore: } \varphi_{max} = 39° \ 25'$$

Time (in min)	$\triangle$ l (in 1/100 mm)	Time (in min)	$\triangle$ l (in 1/100 mm)
1	0.6	17	70.5
2	1.0	18	80.7
3	2.1	19	93.0
4	4.0	20	106.0
5	6.2	21	123.5
6	8.9	22	135.0
7	12.6	23	159.1
8	15.7	24	182.0
9	19.6	25	196
10	23.7	26	216
11	30.0	27	264.0
12	35.1	28	269.0
13	41.2	29	284.0
14	47.5	30	287.0
15	53.7	31	291.0
16	60.0	32	297.0
		33' 30"	

Table 33: Material: Var crushed stone sand; $Q=5$ kg

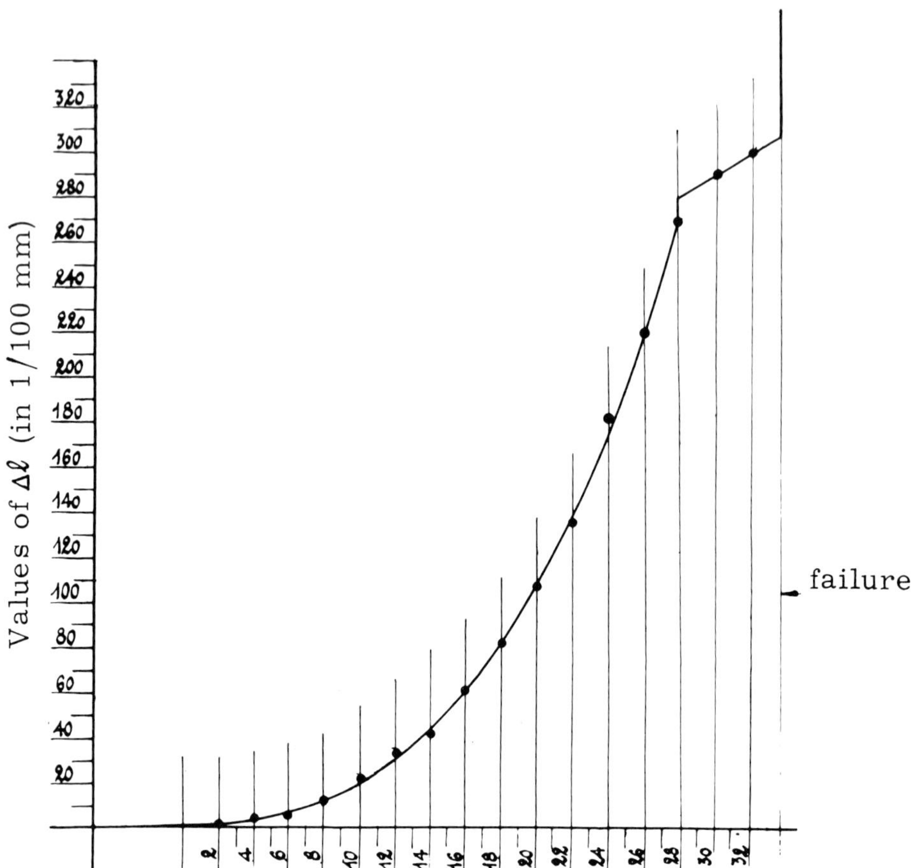

Fig. 73: Material: Var crushed stone sand; $Q = 5$ kg.

Test on Millet with a Load of Q = 5 kg:

The test conditions were the same as in the two preceding cases, the initial load again being $Q = 0.600$ kg (Table 34 and Fig. 74).

As soon as sliding of the plate C has started, it continues even when no further increase is applied to the tensile force. Thus this sliding movement corresponds to final breakdown of equilibrium, i. e., failure.

Load which caused failure: 2.725 kg to which corresponds the minimum coefficient of friction:

$$\text{tg } \varphi_0 = \frac{2.725}{5.000} = 0.545 \quad \text{and therefore:} \quad \varphi_0 = 28° \, 35'$$

Time (in min)	$\triangle$ l (in 1/100 mm)	Time (in min)	$\triangle$ l (in 1/100 mm)
1	1.1	14	29.0
2	1.9	15	34.1
3	2.7	16	44.0
4	3.4	17	56.9
5	4.6	18	72.0
6	5.9	19	92.8
7	8.0	20	117.0
8	10.0	21	142.5
9	12.4	22	175.0
10	15.0	23	214.1
11	18.0	24	246.0
12	21.0	25	350.0
13	24.2	26	failure

Table 34: Material: Millet; $Q = 5$ kg

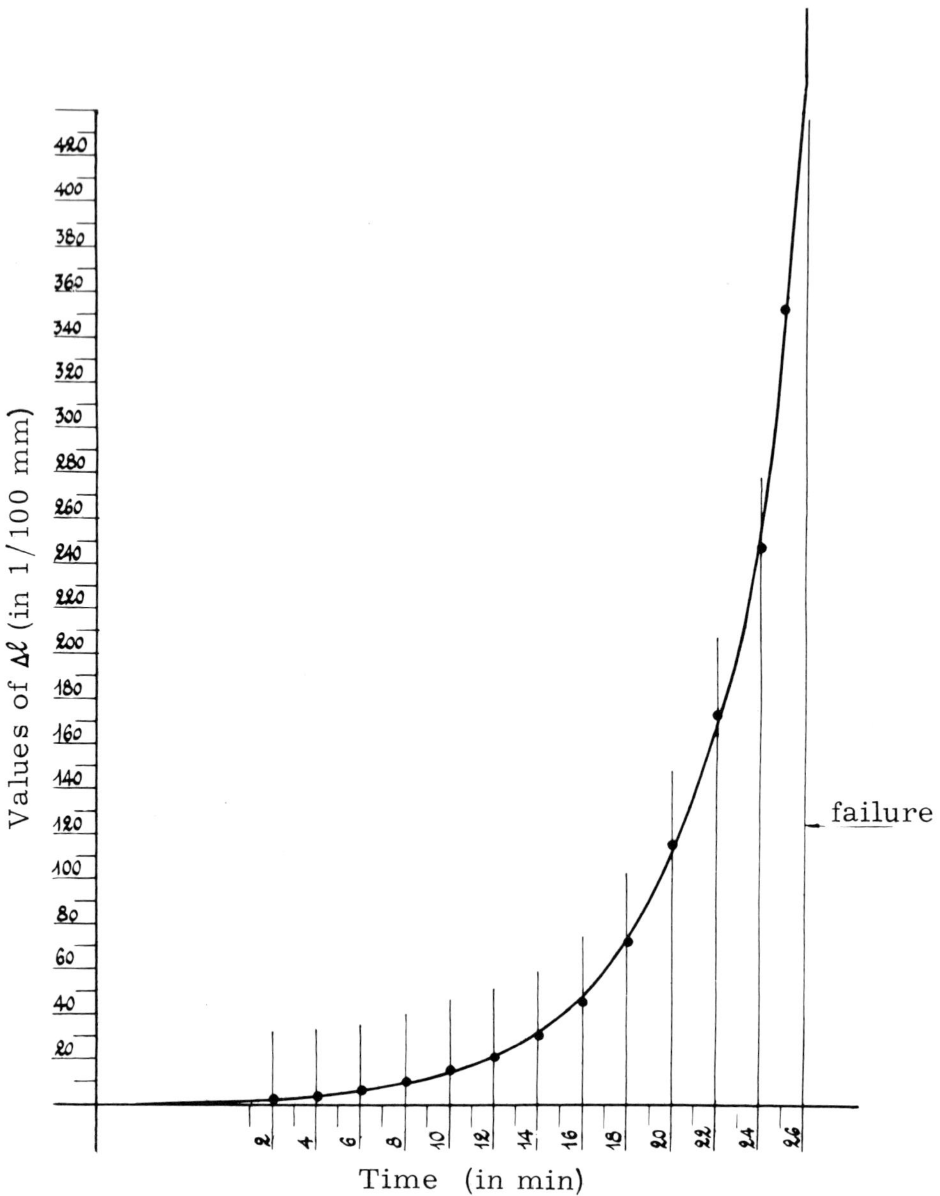

Fig. 74: Material: Millet; $Q = 5$ kg.

Test on Wheat with a Load of $Q=5$ kg:

The test conditions were the same as before, the initial load again being $Q=0.600$ kg (Table 35 and Fig. 75).

Just as in the test performed on millet, as soon as the plate C begins to slide, it continues to move even when no further increase in tensile force is applied. Hence this sliding corresponds to final breakdown of equilibrium, i.e., failure.

The load which caused failure: 3.080 kg to which corresponds the minimum coefficient of friction:

$$\operatorname{tg} \varphi_0 = \frac{3.080}{5.000} = 0.616 \quad \text{and therefore:} \quad \varphi_0 = 31° 35' > \alpha$$

Time (in min)	$\triangle$ 1 (in 1/100 mm)	Time (in min)	$\triangle$ 1 (in 1/100 mm)
1	0.9	13	100.2
2	1.0	14	101.0
3	1.2	15	102.1
4	2.0	16	169.0
5	3.9	17	171.0
6	28.0	18	172.2
7	28.8	19	259.0
8	29.8	20	260.5
9	53.0	21	290.0
10	53.5	22	320.0
11	53.5	23	368.5
12	56.0	24	720
		24' 01"	failure

Table 35: Material: Wheat; $Q=5$ kg

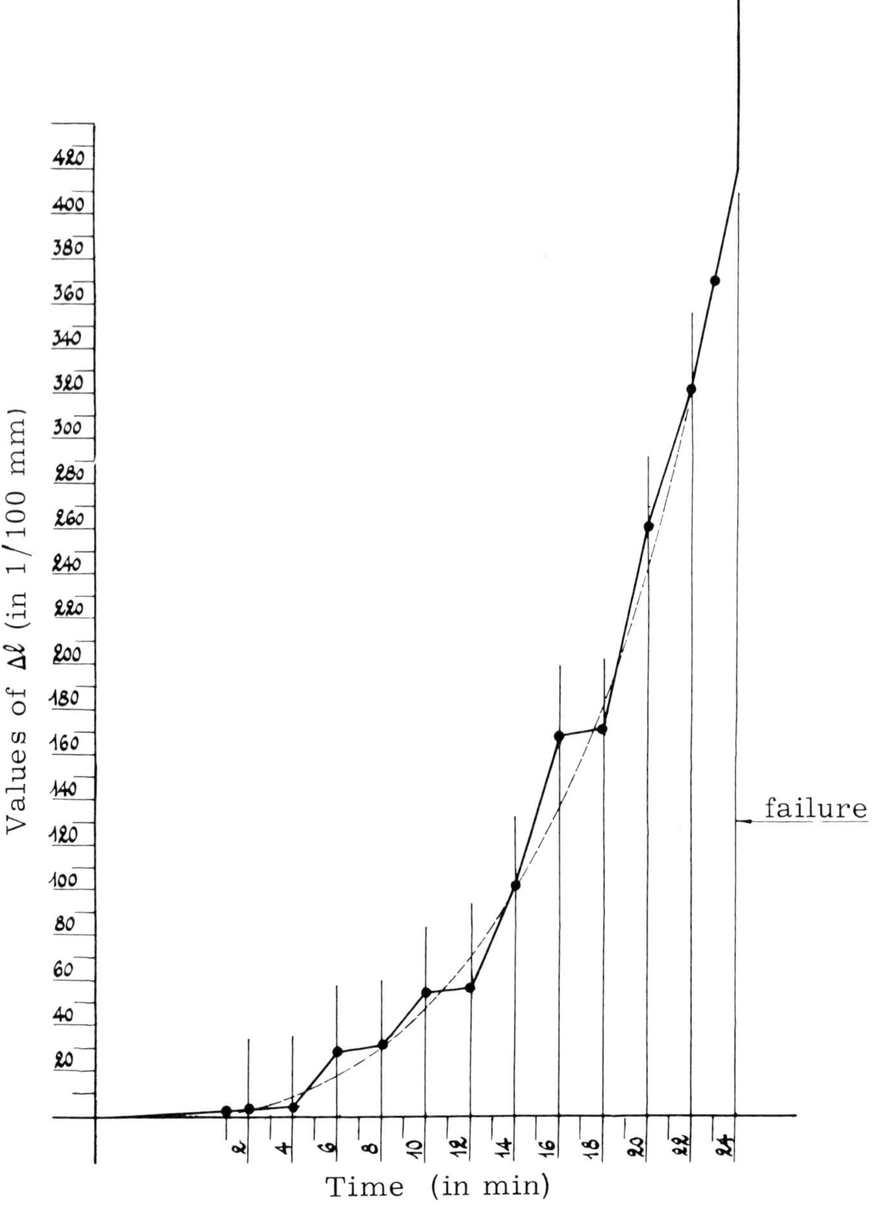

Fig. 75: Material: Wheat; $Q = 5$ kg.

20.3. Settlement of Granular Test Materials

In the tests reported in the foregoing sections the dimensions of the plate C placed on the material were 0.40 m × 0.20 m, and the loads applied to this plate were successively 2.5, 5, 7.5, 10, 12.5 and 15 kg. The corresponding bearing pressures under the plate were therefore:

3.125, 6.25, 9.375, 12.5, 15.675 and 18.75 g/cm².

We have seen that the minimum angle of internal friction decreases with the bearing pressure acting on the material and tends, for zero pressure, to the value of the angle of repose of that material.

The settlement, or "packing", of the material is a function of the pressure, i.e., the compressive stress, due to the load that it is carrying. This settlement, measured for dry fine sand from Garoupe, is given in Table 36.

Applied Loads (in kg)	Cumulative Forces F (in kg)	Settlements (in 1/100 mm)
1.630	1.630	7
+ 0.870	2.500	11.5
+ 2.500	5.000	22.0
+ 2.500	7.500	30.1
+ 2.500	10.000	34.8
+ 2.500	12.500	36.0
+ 2.500	15.000	36.2

Table 36

Weight of the bearing plate: 1.630 kg.

These results have been plotted in Figure 76. We see that, under the test conditions adopted, the settlement tends to a limit value which is soon reached.

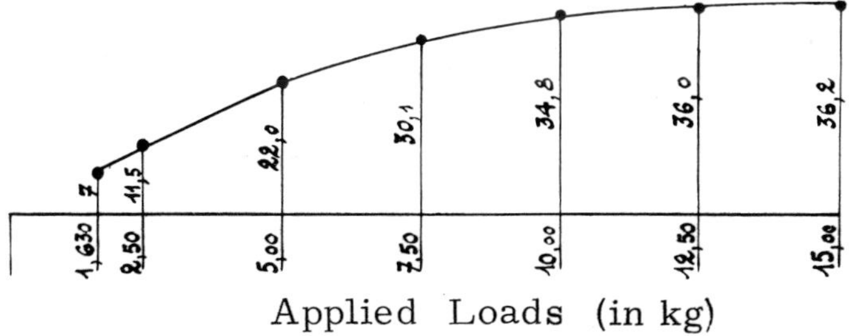

Applied Loads (in kg)

Fig. 76

20.4. Angles of Internal Friction of Wet Fine Sand

For the sake of completeness, the same tests as those previously described were carried out with *wet fine sand* (from the beach at Garoupe-Antibes) with 21%, 24% or 34% (saturated) water content. The results thus obtained showed a marked increase in the minimum and the maximum values of the angle of internal friction in comparison with those obtained with the same sand in the dry state.

20.4.1. Tests on Fine Sand with 24% Natural Moisture

The sand was tested in its natural state, containing 24% water by weight, just as taken from the beach at Garoupe-Antibes.

The test results, for a vertical load of $Q = 5$ kg on the top bearing plate, are given in Table 37 and Figure 77.

Time (in min)	$\triangle$ l (in 1/100 mm)	Time (in min)	$\triangle$ l (in 1/100 mm)
	1.3	8	74.7
1	4.9	9	95.7
2	8.8	10	122.0
3	14.8	11	150.5
4	22.5	12	190.0
5	31,9	13	252.0
6	43.2	14	345.0
7	56,8	14 min 45 s	1st sliding

Table 37

The load which caused first sliding movement: 4.760 kg to which corresponds the minimum angle of friction:

$$\text{tg } \varphi_0 = \frac{4.760}{5,000} = 0.952 \quad \text{and therefore:} \quad \varphi_0 = 43° \, 38'$$

The load which caused failure: 5.100 kg to which corresponds the maximum angle of friction:

$$\text{tg } \varphi_{max} = \frac{5.100}{5.000} = 1,02 \quad \text{and therefore:} \quad \varphi_{max} = 45° \, 40'$$

400
380
360
340
320
300
280
260
240
220
200
180
160
140
120
100
80
60
40
20
Values of Δℓ (in 1/100 mm)
1 2 3 4 5 6 7 8 9 10 11 12 13 14
Time (in min)

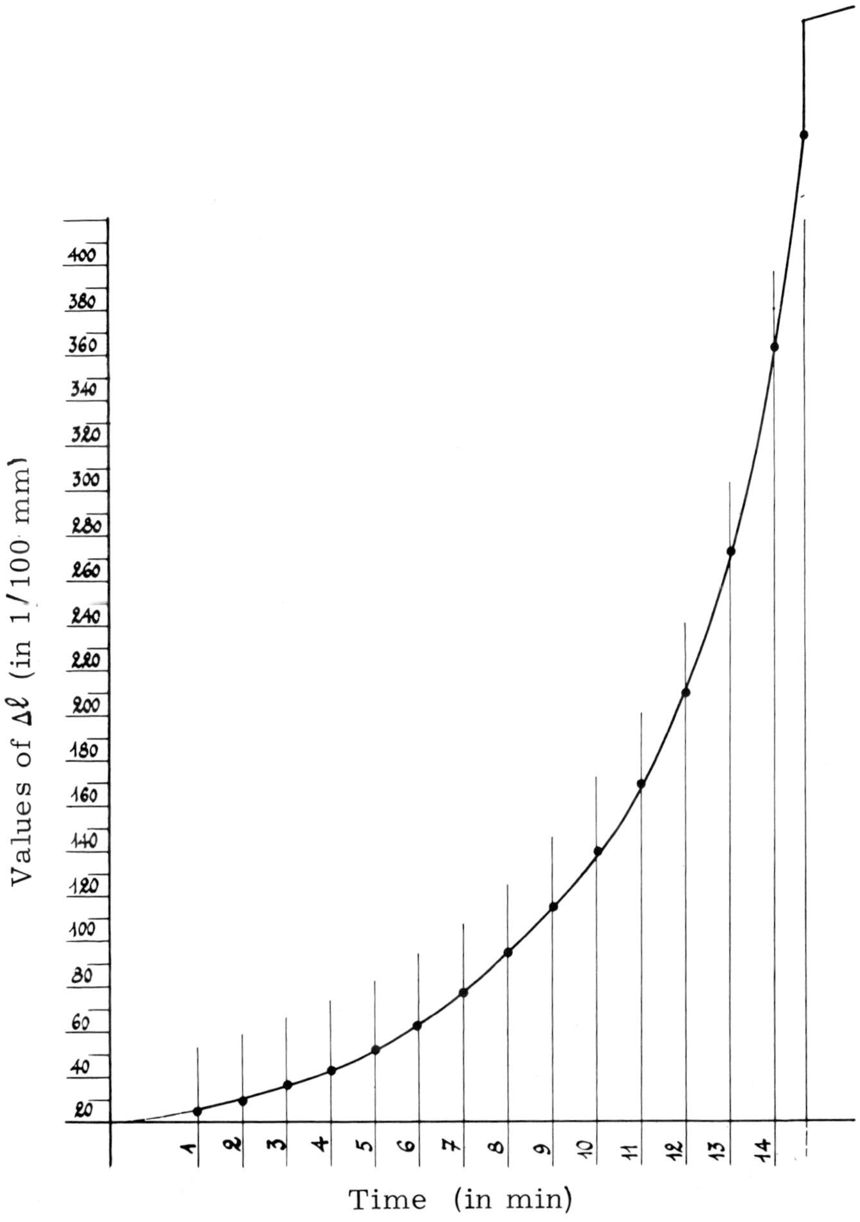

Fig. 77

20.4.2. Tests on Fine Sand with 21% Moisture

The wet fine sand with a water content of 24% by weight, as taken from the beach at Garoupe-Antibes, was spread on a porous slab taking the place of the plate A in the apparatus illustrated in Figure 69.

As a result, the moisture content was lowered to 21%.

The test results, for a vertical load of $Q=5$ kg on the top bearing plate, are given in Table 38 and Figure 78.

The load which caused first sliding movement: 4.290 kg to which corresponds the minimum angle of friction:

$$\text{tg } \varphi_0 = \frac{4.290}{5.000} = 0.858 \quad \text{and therefore:} \quad \varphi_0 = 41° \ 40'$$

The load which caused failure: 4.620 kg to which corresponds the maximum angle of friction:

$$\text{tg } \varphi_{max} = \frac{4.620}{5.000} = 0.924 \quad \text{and therefore:} \quad \varphi_{max} = 42° \ 45'$$

Time (in min)	$\triangle\, l$ (in ·1/100 mm)
	0.2
1	1.0
2	3.8
3	9.1
4	15.9
5	24.0
6	32.8
7	43.0
8	54.8
9	69.4
10	86,7
11	109.0
12	131.0
13	157.8
14	193.5
15	243.3
16	293.5
17	395.0
	890.0
18 min 35 s	failure

Table 38

900
880
860
840
38
36
34
32
30
28
26
24
22
20
18
16
14
12
10
8
6
4
2
Values of Δl (in 1/100 mm)
1
2
3
4
5
6
7
8
9
10
11
12
13
14
15
16
17
First
Sliding
failure
Time (in min)

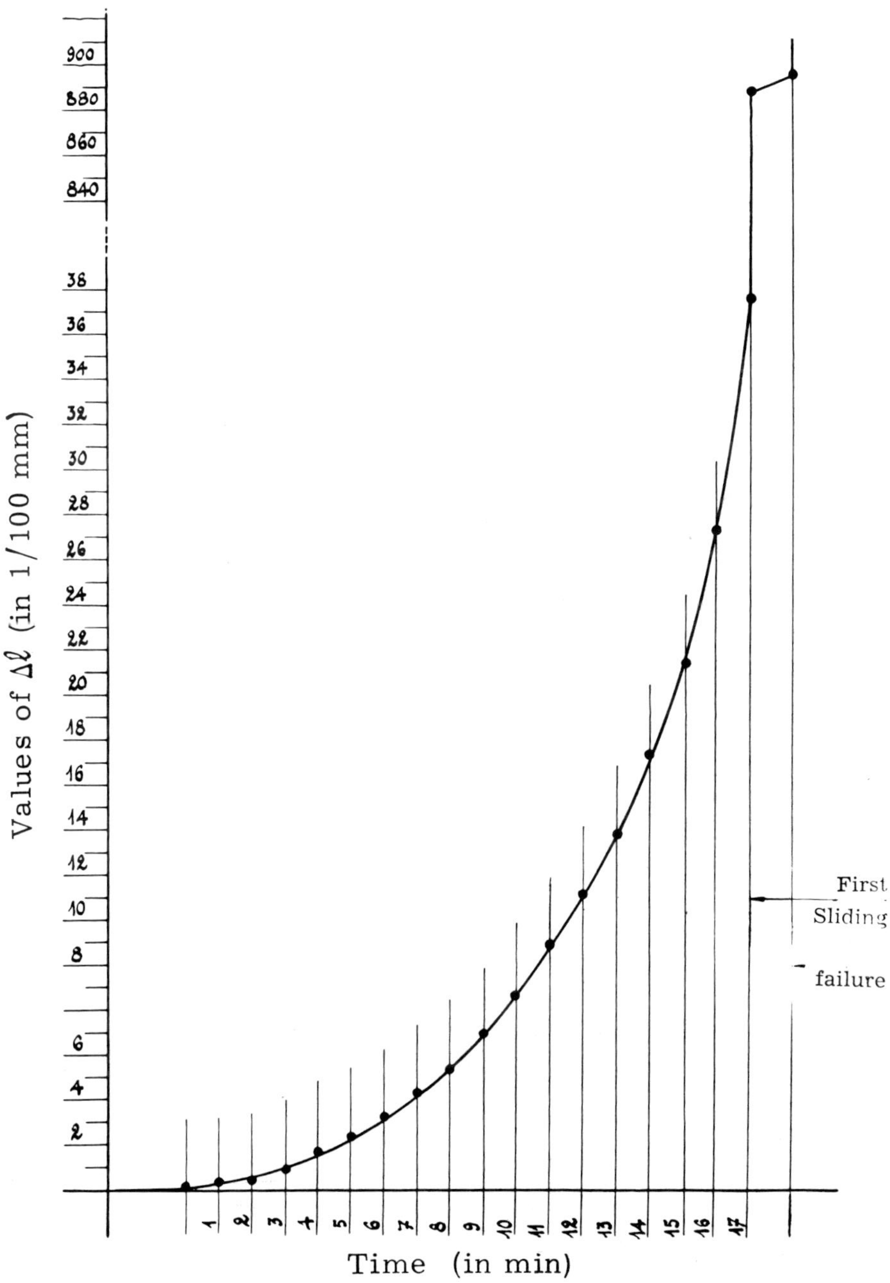

Fig. 78

20.4.3. Tests on Saturated Fine Sand

The test equipment employed was the same as described in Section 20.1., except that now the plate A (Fig. 69) was replaced by a parallelepipedal tank on the bottom of which the sand for testing was placed. Water was poured into the tank so as to immerse the sand completely up to the level of the underside of the plate C carrying the load Q.

The test results obtained for $Q = 5$ kg are given in Table 39 and Figure 79.

Time (in min)	$\triangle\, l$ (in 1/100 mm)
	3.5
1	5.2
2	7.4
3	10.0
4	12.9
5	15.8
6	18.9
7	22.3
8	25.7
9	29.8
10	34.6
11	39.7
12	45.0
13	50.8
14	57.8
15	67.1
16	78.5
17	90.9
18	106.4
19	126.0
20	149.2
21	174.6
22	204.8
23	243.5
24	283.5
24 min 50 s	sliding

Table 39

The load which caused first sliding movement (without stabilisation, contrary to the preceding tests): 4.140 kg to which corresponds the angle of friction:

$$\operatorname{tg} \varphi = \frac{4.140}{5.000} = 0.828 \quad \text{and therefore:} \quad \varphi = 39° \, 35'$$

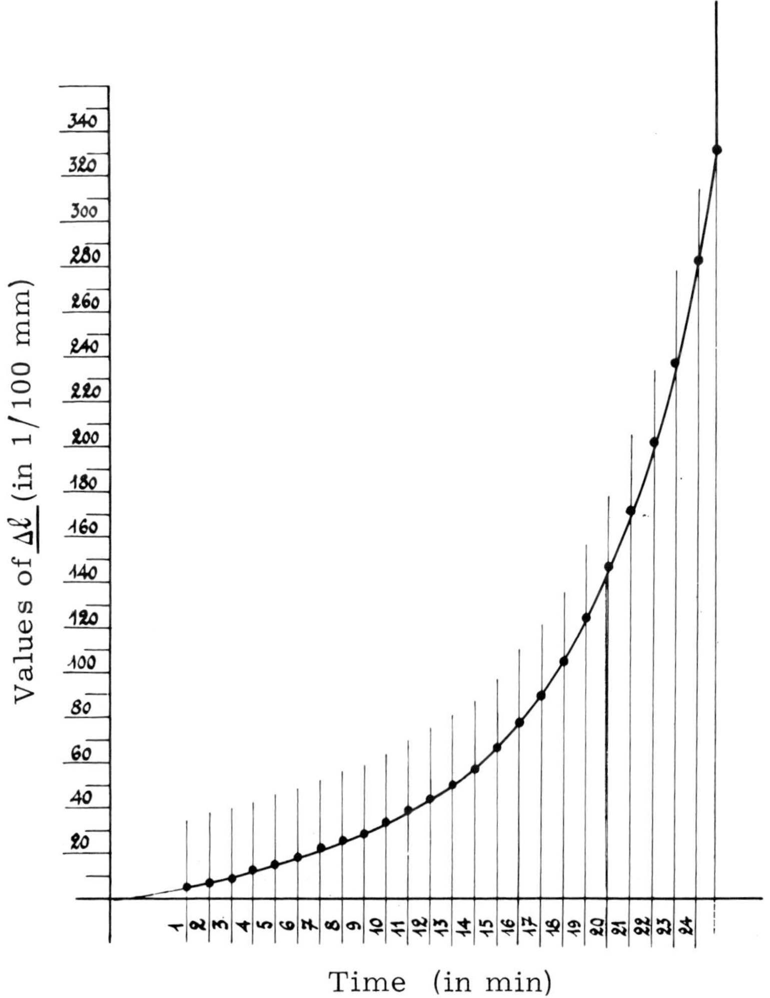

Fig. 79

20.5. Practical Consequences for Retaining Wall Design

The tests were performed with a top bearing plate C with dimensions of 0.20 m × 0.40 m subjected to vertical loads of $Q=2.5$, 5, 10 and 15 kg (including the weight of the plate itself). The corresponding pressures (compressive stresses) on the test material were 3.125, 6.25, 12.5 and 18.75 g/cm^2 respectively. We see that the minimum angle of internal friction increases with the compressive stresses and thus with the "interlock" developed between the grains.

Since the maximum compressive stress applied in the tests (18.75 g/cm^2) corresponds to a height of sand of about 13.5 cm, we see that the actual compressive stress of the grains within the mass of material is evidently greater than the maximum pressure applied to the top surface thereof and that the minimum angle of internal friction is thus at least $35°$ while the angle of repose is $33°40'$.

If we calculate the thrust exerted on a vertical retaining wall of height h by sand such as the fine sand used in these tests, by applying the formula expressing this thrust as a function of the angle of repose, we obtain:

$$P = \frac{\gamma \cdot h^2}{2} \cdot \left(\frac{\pi - 2 \times 33°\ 40'}{\pi + 2 \times 33°\ 40'} \right)^2 = 0.208 \cdot \frac{\gamma \cdot h^2}{2}$$

whereas this same formula expressing the thrust as a function of the minimum angle of internal friction gives:

$$P = \frac{\gamma \cdot h^2}{2} \cdot \left(\frac{\pi - 2 \times 35°}{\pi + 2 \times 35°} \right) = 0.1935 \cdot \frac{\gamma \cdot h^2}{2}$$

The difference between the two results is therefore about 7.5%.

Hence we see that when a retaining wall is designed by applying the proposed formulas with the angle of repose instead of the minimum angle of internal friction we shall be on the safe side.

20.6. Comparison of the Results Obtained by the Friction Plate Method and with the Triaxial Compression Apparatus

It has been shown that the angle of repose of fine sand is $\alpha = 33°40'$ and that the minimum angle of internal friction obtained by the friction plate method has the following values:

34°02′ for a compressive stress of 3.125 g/cm²
34°20′ for a compressive stress of 6.250 g/cm²
34°50′ for a compressive stress of 12.500 g/cm²
35°   for a compressive stress of 18.750 g/cm²

In the laboratories of Bureau Véritas triaxial compression tests were performed on the same sand with a view to determining the minimum angle of internal friction thereof. The results obtained are given in Table 40.

	Total hydrostatic pressure (bars)	Rate of vertical deformation (mm/min)	Maximum stress (bars)	Value of φ
		0.5	6.522	41° 50′
Test No. 1	0.1	1.0	9.024	41° 47′
		1.5	11.522	
		2.0	13.920	40° 57′
		0.2	1.393	40° 42′
Test No. 2	0.4	0.4	2.341	40° 42′
		0.6	3.292	
		0.8	4.280	41° 32′
Test No. 3	0.4	1.0	5.025	41° 47′
		2.0	10.000	

Table 40

We see that, irrespective of the rate of vertical deformation of the specimen used in the test and irrespective of the hydrostatic pressure to which it is subjected, triaxial compression testing gives values for angle of internal friction that are distinctly higher than those obtained by the friction plate method.

20.7. Conclusions

The following conclusions can be drawn from the experiments reported.

1. In the case of sands the deformation curves reveal the values of a minimum coefficient of friction and a maximum coefficient of friction.

2. The minimum angle of internal friction is a little larger than the angle of repose. The experiments showed this minimum angle to decrease with the magnitude of the stress applied. For zero stress it would in fact be equal to the angle of repose.

3. It is therefore permissible to assume, for the design of retaining structures, that in the limit case:

$$\varphi_0 \geqslant \alpha$$

20.8. Important Final Remark to Section 20

The expression for the deformation of a mass of granular material loaded so as to develop passive resistance is of the following form:

$$\Delta \ell = h \cdot \chi \, (k_p - 1)$$

According to this formula, it is indeed apparent that if a higher value for the passive resistance coefficient were taken into account, this would involve taking account of a larger amount of deformation, of such magnitude as to entail the risk of structural disruption if this deformation were incompatible with the nature of the structure concerned – for example, in the case of a diaphragm wall.

So it is necessary to determine the value of the coefficient of internal friction φ with the greatest possible accuracy, in so far as calculations relating specifically to passive resistance phenomena are concerned.

On the other hand, in the case of active pressure (thrust) phenomena, for which it is the maximum values which must more particularly be adopted in the calculations, the magnitude of the thrust should be estimated – on the safe side – essentially in terms of the value of the angle of repose α of the retained material. This angle can quite easily be determined without ambiguity.

21. Appendix

Theoretical Formulation in Confirmation of the Underlying Principles in Defining the Equilibrium of a Cohesionless Granular Material Retained by a Wall.

The formulas for thrust and for passive resistance, with which Volumes I and II are concerned, define the whole mechanism of the equilibrium of a granular material retained by a wall.

The formulas in Volume I relate to the "at rest" state of equilibrium of such materials. They all were established by a rigorous interpretation of the experimental results and they provide every guarantee of safety and stability of the structures in the design of which they are applied.

Experiments relating to the equilibrium of granular materials, beyond the "at rest" state of equilibrium thereof, are dealt with in Volume II. Although they are more particularly concerned with passive resistance – or passive pressure – in the case of deformation of the materials in question, these experiments also provide instructive confirmation of the formulas for "at rest" conditions established in Volume I, and the definitive information derived therefrom would have enabled the conclusions from the whole investigation to be finalised.

However, the various thrust coefficients and passive resistance coefficients that have been established present such a degree of harmony among one another – which is given specific expression by their incorporation into a universal formula – that there was every reason to suppose that a theory could be deduced therefrom which could strengthen the experimentally gained knowledge of the mechanism of the equilibrium of granular materials.

The present Appendix is concerned with this theoretical formulation, the significance of which lies in the confirmation of the remarkable agreement between theory and experiment.

A.1. Analytical Investigation of the Value of the Dihedral Angle of the Active Pressure Wedge and of the Thrust Coefficient

The value of the dihedral angle of the wedge, or prism, associated with failure due to active pressure was determined experimentally, on a first occasion, by interpretation of the results of tests which enabled the thrust coefficient – in the case of granular material with horizontal top surface and retained by a vertical wall – to be defined in terms of the angle of repose α, taken as equal to the minimum angle of internal friction φ of the granular material considered:

$$k_a = \left(\frac{\pi - 2\,\alpha}{\pi + 2\,\alpha}\right)^2$$

Now this coefficient is, with fair approximation, equivalent to (see page 97 of Vol. I):

$$k_a = \text{tg}^4\left(\frac{\pi}{4} - \frac{\alpha}{3}\right)$$

Hence the phenomena in question manifest themselves as though the angle of the failure wedge were:

$$\beta = \frac{\pi}{4} - \frac{\alpha}{3}$$

A large number of check experiments were performed with granular materials having an inclined or a horizontal top surface, with or without surcharge, which confirmed the constant value of this angle.

So there were grounds for seeking a theoretical explanation for this observed fact.

A.1.1. Recapitulation

For determining the value of the angle β of the thrust wedge (active pressure wedge) it is relevant first to recall the known facts and the following actual phenomena previously established by observation:

– the thrust, i.e., the resultant force due to the active pressure exerted by the retained material, has its maximum value in the "at rest" state of equilibrium of the material (see Volume I);

– this maximum value can be expressed in terms of the minimum angle of internal friction φ_0, which can be taken as being equal to the angle of repose α of the granular material:

– according to Résal (11):
"The calculations should be based on the angle of repose corresponding to the most unfavourable conditions that appear liable to arise."

– according to Terzaghi and Peck (8):
"The lowest value of the angle of internal friction φ is equal to the angle of repose."

– the value of the dihedral angle β of the thrust wedge is constant from the "at rest" state of equilibrium of the material up to failure of that equilibrium, no matter how much the retaining wall is tilted, what the top surface slope of the material is (horizontal or inclined), whether or not there is a surcharge, and whether the rear surface of the wall is smooth or rough (see Volume I, Chapter: Observation of the sliding of the failure wedge; photographs 76 to 101);

– the thrust is not perpendicular to the rear of the wall (Fig. A 1). According to Mayer (9):

"Its direction is known because it forms, with the normal to the face of the wall, an angle equal to the angle of friction of the material retained by the wall".

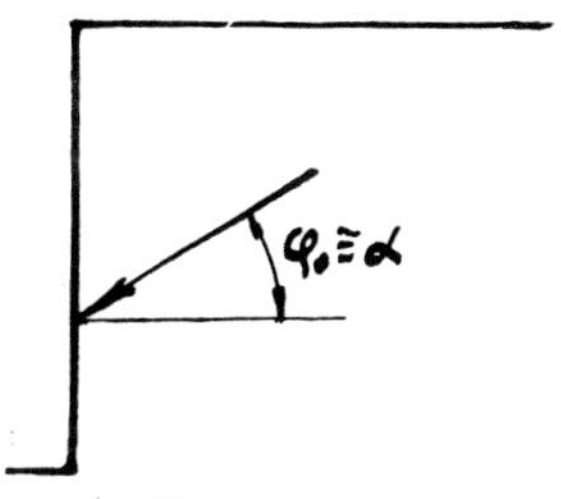

Fig. A 1

Basing themselves on these data, the formulas which will be established in the present Appendix are therefore expressed in terms of the angle of internal friction φ of the granular material (for which the minimum value φ_0 is adopted) and, with equal validity, in terms of the angle of repose α, since these respective angles are equivalent.

This significant point, namely:

$$\varphi_0 \cong \alpha$$

will be highlighted a number of times in this treatment of the subject, in order to preclude any possible confusion, for it must be emphasised that our research is directed, first and foremost, at determining the maximum thrust which physically occurs at the minimum value of the angle of internal friction of the material.

A.1.2. Observation of the Actual Phenomena Associated with the Formation of the Thrust Wedge

The formation of a mass of granular material with a naturally sloped top surface and retained by a wall can be conceived as taking place in two successive stages:

– First, the material attains a state of equilibrium, with its top surface AM inclined at the angle of repose α of that material and passing through the base A of the retaining wall (Fig. A2).

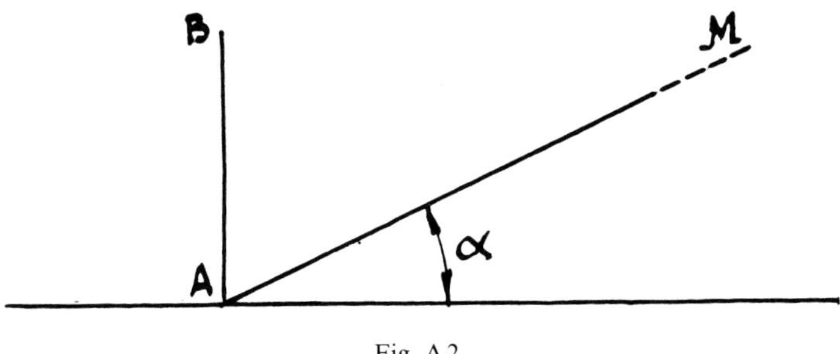

Fig. A 2

– Next, when the volume of material behind the wall is further increased, the wall will constitute an obstacle preventing the grains from spreading freely beyond A (Fig. A3). Thus a discontinuity is formed in the mass of material. This discontinuity produces a particular state of equilibrium characterised by the formation, within the mass and against the rear face of the wall, of the thrust wedge ABC which Coulomb defined for the case where the top surface of the material is horizontal.

It is therefore necessary to examine the phenomenon due to the formation of this wedge in the case of granular materials in order to gain an understanding of the mechanism whereby the actual value of the angle BAC of this wedge can be determined.

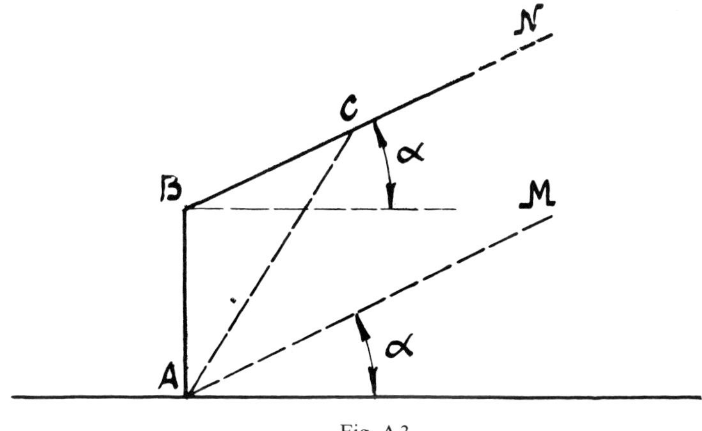

Fig. A 3

A.1.3. Examination of the Formation of the Thrust Wedge in the Case of Cohesionless Granular Material Retained by a Wall

Close observation of the movement of the grains of the granular material retained by a wall which swivels about its base provides the confirmation and the explanation of the phenomenon of the formation of the thrust wedge. Inside the mass of material this wedge is bounded by a sliding plane whose slope with respect to the vertical is different from the angle of repose that this same material would adopt by itself, in the absence of any retaining wall.

A.1.3.1. Material with Horizontal Top Surface

Consider an element m at the horizontal top surface of a mass of granular material retained by a vertical wall. If the latter is tilted through an angle θ of arbitrary magnitude by letting it swivel about its base A, the element m is found to be involved in the observed deformation of the mass, under the double effect due to the action of the dead weight of the element (in the vertical direction) and to the sliding of the element in the direction of the slope along which the granular material flows.

The characteristic directions of displacement of the element are represented schematically by the respective arrows in Fig. A4a.

The element m thus moves to m′, and the angle that the line mm′ forms with the vertical is equal to the dihedral angle of the thrust wedge. In fact, all the experiments show (photographs 76 to 101 in Volume I) that the paths of the grains of the material are substantially parallel to the general sliding plane of the wedge.

Besides, it can readily be verified that an element located at point B at the top of the retaining wall will, when the wall has tilted, be at point B' located on the line drawn through B and parallel to the general sliding plane AC. This phenomenon is perfectly explicable in that the volume of the thrust wedge is constant (except for the decompression of the material attendant upon the forward tilting of the wall) and that the points B and B' are thus both at the same distance from the base AC. The cogency of this reasoning is borne out by the experiment to which Fig. 80 in Volume I relates.

With reference to that experiment it is to be noted that the displacement behaviour of any particular element m within the mass of material is the same as that of the element m at the surface thereof, as envisaged above.

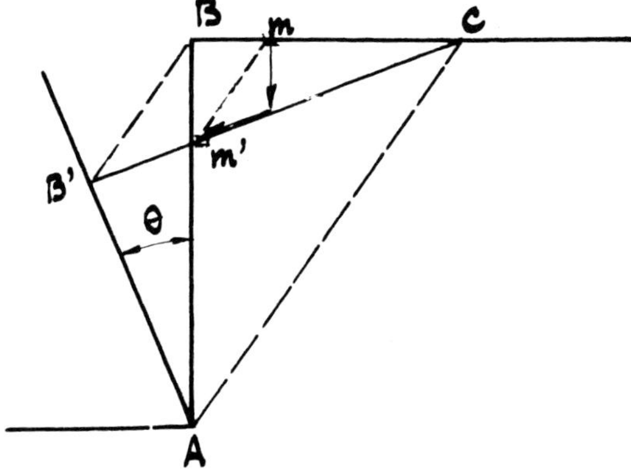

Fig. A4a

A.1.3.2. Material with Inclined Top Surface

When the wall swivels about its base A and thus moves to the position AB' (Fig. A4b), the material will flow forward naturally behind it. This flow will occur in layers sloped at the angle of repose α.

As a result, the top surface of the material, initially corresponding to BN, will go down to DN', when the retained material attains its new state of equilibrium.

This equilibrium, however, can be achieved only after an intermediate stage characterised by movement of the grains of the material, below the plane corresponding to DN' and in the vicinity of the wall, so as to enable the material to adjust itself to the new position of the wall, as in the preceding case.

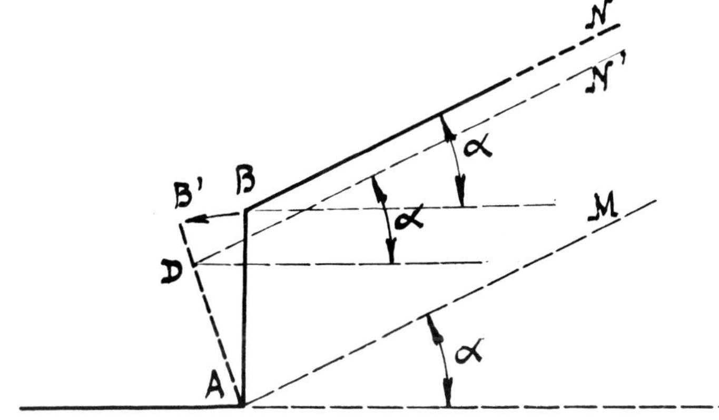

Fig. A4b

Consider an element m, of weight p, within the wedge (Fig. A5). When the wall tilts forward, from AB to AB' this element will – as in the preceding case follow a direction r intermediate between the vertical (in which direction the element tends to travel under the action of gravity) and the line f sloping at an angle α (in which direction the element tends to travel because of the natural sliding movement of the grains). Numerous experiments carried out by various investigators and also by the present authors (as described in Volume I) have shown that – just as in the case of material with horizontal top surface – the direction of the resultant movement r of the grains inside the thrust wedge in material whose top surface is inclined

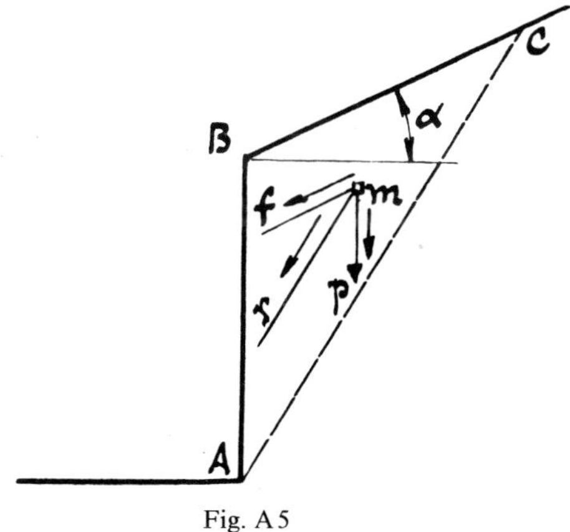

Fig. A5

is, in the state of equilibrium failure, constant and parallel to the sliding plane
AC of the wedge. This is to be observed up to the final stage of equilibrium
failure of that wedge following a movement (even a small one) of the wall,
and it continues to be so when the movement of the wall becomes greater,
as is apparent from the photographs 76 to 101 in Volume I and from the
comments accompanying them. This general sliding plane then forms an
angle β with the vertical, as can be inferred from the foregoing remarks
based on the observation of the actual phenomena and from a line of
reasoning which is valid in all cases of retained granular material – with
horizontal or inclined top surface and with vertical or inclined retaining
wall – as a generalisation not reported by Coulomb in his paper dealing
with the subject.

A.1.3.3. Important Remark

In Sections A.1.3.1. and A.1.3.2. the reference to the substantial tilting of
the retaining wall has been made – as was already done with regard to the
experiments to which the photographs 76 to 101 in Volume I refer – only
with a view to highlighting the phenomenon of adaptation of the loose
granular material.

But in reality this phenomenon is continuous from the very outset, pro-
ceeding directly from the potential elements of the material at rest in the
vicinity of the wall retaining them, up to the limit of the thrust wedge defined
by the sliding plane thereof.

Under these conditions the analysis presented in the following Section
refers, along the wall, to the relative limit of equilibrium at rest and can
be regarded as arising from this equilibrium which produces the maximum
thrust with which alone we are concerned. In actual fact even a very small
displacement of the retaining wall immediately brings about a decrease in
the thrust.

A.1.4. Determination of the Dihedral Angle of the Thrust Wedge

The magnitude of the angle β of the thrust wedge is directly determinable
from an investigation of the equilibrium of this wedge under the action of
the forces to which it is subjected:

1. The weight of the thrust wedge (Fig. A6a):

$$P = \frac{\gamma \cdot H^2}{2} \cdot \text{tg}\,\beta$$

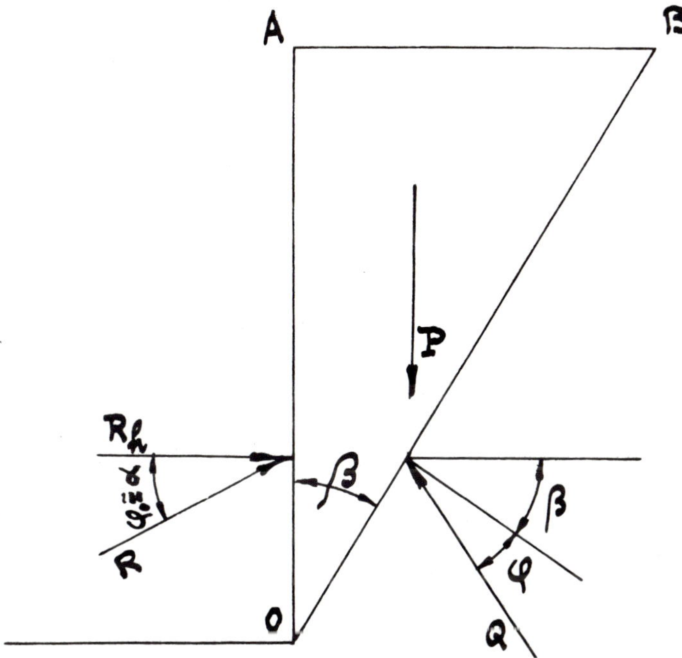

Fig. A6a

2. The resultant R of the reaction of the wall in response to the thrust, in the case of material with horizontal top surface and retained by a vertical wall.

   This resultant has its maximum value when its slope is equal to the minimum angle of internal friction φ_0 (which is to be taken as equal to the angle of repose α (see Section A.1.1.).

3. The resultant Q of the reaction of the material tending to oppose the movement of the wedge along the sliding plane OB.

Taking the sums of their projections, the equilibrium conditions of the forces can be written as follows:

$$P = \frac{\gamma \cdot H^2}{2} \cdot \operatorname{tg} \beta = Q \sin (\beta + \varphi) + R \sin \alpha$$

$$R \cos \alpha = Q \cos (\beta + \varphi)$$

On eliminating Q from these two equations, we obtain the value of the horizontal component R_n of the thrust:

$$R_h = R \cos \alpha = \frac{\dfrac{\gamma \cdot H^2}{2} \cdot \text{tg}\,\beta}{\text{tg}\,\alpha + \text{tg}\,(\beta + \varphi)}$$

or:

$$R_h = \frac{\gamma \cdot H^2}{2} \cdot \frac{\text{tg}\,\beta}{\text{tg}\,\alpha + \text{tg}\,(\beta + \varphi)}$$

whence we find for the thrust coefficient:

$$k_a = \frac{\text{tg}\,\beta}{\text{tg}\,\alpha + \text{tg}\,(\beta + \varphi)}$$

But we do not know the value of β whose maximum was determined by Coulomb by finding the derivative of the above expression and equating it to zero: $dk_a/d\beta = 0$. He simplified the problem of solving this equation by assuming zero inclination of the thrust (i.e., $\alpha = 0$), so that his expression for the thrust coefficient became:

$$k_a = \frac{\text{tg}\,\beta}{\text{tg}\,(\beta + \varphi)}$$

By determining the derivative of this expression, the value of the angle β is found (see Kérisel (12)):

$$\beta = \frac{\pi}{4} - \frac{\varphi}{2} \qquad (1)$$

and therefore we obtain for the thrust coefficient:

$$k_a = \text{tg}^2 \left(\frac{\pi}{4} - \frac{\varphi}{2} \right)$$

In this way Coulomb arrived at the conventional thrust coefficient, based on his assuming a simplifying hypothesis which the present authors have here amplified.

So we must go back to the facts as they really are and take account of the actual value of the inclination of the thrust, which considerably alters the angle $\pi/4 - \varphi/2$, as numerous experiments have proved.

In fact, the photographs 76 to 101 in Vol. I, already referred to, provided indisputable confirmation that throughout the deformation of the retained material – corresponding to displacement of the wall by swivelling about its base, which occurs at substantially constant volume (except for the volume change due to decompression of the thrust wedge) – each element thereof slides in a direction practically parallel to the general sliding plane of the wedge.

Now the mass of granular material is composed of separate elements (grains) which interact. It is therefore permissible to analyse the overall phenomenon of the deformation of the wedge along the general sliding plane by starting from an examination of the equilibrium of one element m of that wedge.

For that purpose, while adopting Coulomb's theoretical value, on the only assumption made by him – without experimental confirmation – namely, zero inclination, we can represent the action of the element m by the vector F (Fig. A6b), whose direction is inclined at the angle $\pi/4 - \varphi/2$ in relation to the vertical.

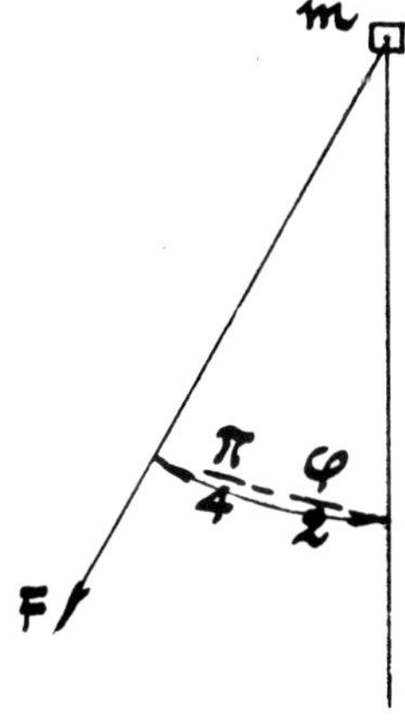

Fig. A6b

With the known direction of the vector F and its vertical component equal to the weight V of the element m, we obtain for the horizontal component (Fig. A6c):

$$H = V \, \text{tg} \left(\frac{\pi}{4} - \frac{\varphi}{2} \right)$$

Inside the thrust wedge the element m is pulled vertically downwards under the action of its own weight.

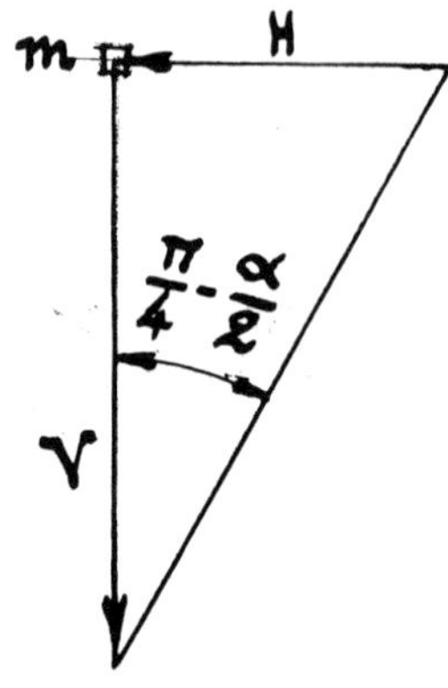

Fig. A6c

But, under the action of the potential elements referred to in Section A.1.3.3., e.g., the action tending to make the grains slide in a direction corresponding to the angle of repose, a mobilisation of the friction occurs, the effect of which upon the general phenomenon of the equilibrium of the thrust wedge was not examined by Coulomb. Here, then, is a gap that ought to be filled. According to Coulomb's law this mobilisation of the friction will produce a tangential action H' by the force H upon the plane perpendicular to its direction:

$$H' = H \, tg \, \varphi$$

or (see A. 1. 1.):

$$H' = H \, tg \, \alpha$$

and hence:

$$H' = V \, tg \left(\frac{\pi}{4} - \frac{\alpha}{2} \right) tg \, \alpha$$

H' is represented by OO' in Fig. A7.

Under the action of H and H' the point O moves to v, while V undergoes translatory displacement from V to V', represented by vv'.

But, correlatively, the horizontal sliding movement under the action of H means that, in the same manner as before, Coulomb's law of friction is applied to the vector V, which has moved to V', and perpendicularly to the direction thereof (Fig. A7):

$$V'' = V' \, tg \, \alpha = V \, tg \, \alpha$$

V'' is represented by $v'v''$ in Fig. A7.

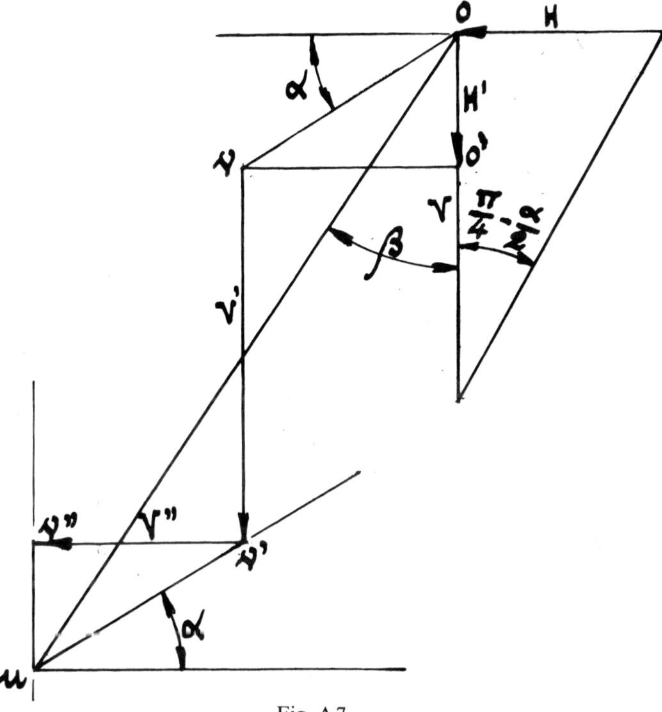

Fig. A 7

The movement in the direction "f", as referred to in Section A.1.3.2. (Fig. A5), is represented by the projection of V'' on the plane of repose, parallel to the direction of the force represented by the vector V', at a point u as indicated in Fig. A7.

From this analysis of the phenomenon with regard to the two loading conditions to which the various elements of the thrust wedge are subjected it follows that the line ou, passing through the original position o of the element before any deformation occurred and through the point u to which the element moved after equilibrium failure of the material, must represent the resultant direction of movement of the elements m of the thrust wedge.

So we find that the direction ou corresponding to the movement associated with the phenomenon of deformation of the thrust wedge is indeed inclined relatively to the vertical at an angle different from, and larger than, $\pi/4 - \alpha/2$.

We must therefore calculate the value of that angle in terms of the data given above, in order to compare it with the value determined experimentally by interpretation of the test results.

A.1.5. Trigonometric Expression of the Value of the Angle β

Consider an element m of the thrust wedge and assume this element to have a weight V of unit value, i.e., $V=1$ (Fig. A8).

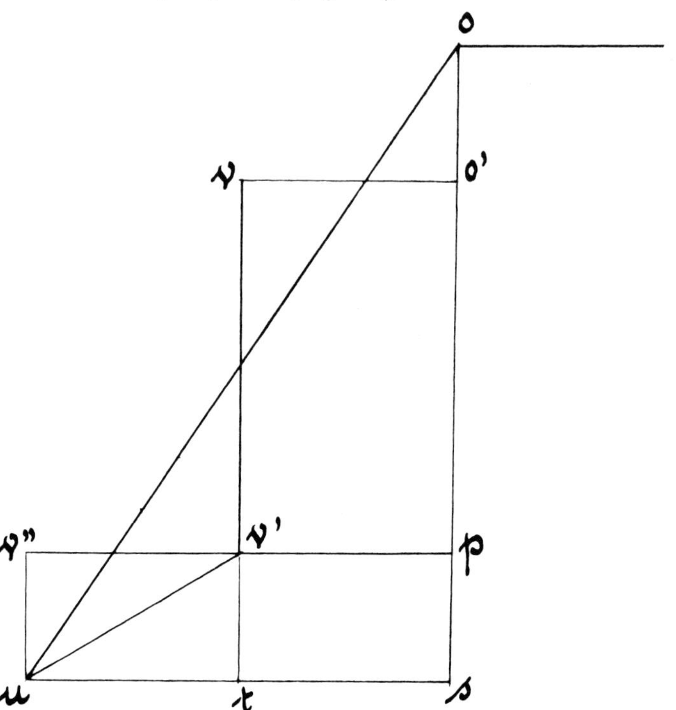

Fig. A8

The value of the angle β is then obtained from the following calculation based on the principles set forth in Section A.1.4.:

$$oo' = \mathrm{tg}\left(\frac{\pi}{4} - \frac{\alpha}{2}\right)\mathrm{tg}\,\alpha$$

$$o'p = 1$$

$$o'v = ts = \mathrm{tg}\,\alpha$$

$$v'v'' = ut = \mathrm{tg}\,\alpha$$

$$v''u = ps = v'v''\,\mathrm{tg}\,\alpha = \mathrm{tg}^2\,\alpha$$

$$us = ut + ts = \mathrm{tg}\,\alpha + \mathrm{tg}\left(\frac{\pi}{4} - \frac{\alpha}{2}\right)$$

$$os = oo' + o'p + ps = \mathrm{tg}\left(\frac{\pi}{4} - \frac{\alpha}{2}\right)\mathrm{tg}\,\alpha + 1 + \mathrm{tg}^2\,\alpha$$

And we finally obtain:

$$\boxed{\; \mathrm{tg}\,\beta = \frac{\mathrm{tg}\,\alpha + \mathrm{tg}\left(\dfrac{\pi}{4} - \dfrac{\alpha}{2}\right)}{1 + \mathrm{tg}\,\alpha\;\mathrm{tg}\left(\dfrac{\pi}{4} - \dfrac{\alpha}{2}\right) + \mathrm{tg}^2\,\alpha} \;}$$

Formula (1)

We can then compare the value of the angle β calculated from the above formula (1) with the value obtained experimentally by interpretation of the test results:

$$\beta = \frac{\pi}{4} - \frac{\alpha}{3}$$

Angle	Values of β	
$\alpha \cong \varphi$	Formula (1)	$\pi/4 - \alpha/3$
0	45°	45°
24°	36°20′	37°
27°	35°30′	36°
30°	34°40′	35°
33°	33°55′	34°
36°	33°05′	33°
39°	32°10′	32°
45°	30°20′	30°

Table 41

A.1.6. Conclusions from Sections A.1.3. to A.1.5.

It follows from Table 41 that the value of the angle β of the thrust wedge is practically equal to:

$$\beta = \frac{\pi}{4} - \frac{\alpha}{3}$$

and that this is generally so, because the reasoning on which it is based is equally applicable to any granular cohesionless material with horizontal or inclined top surface and irrespective of whether the wall that retains it is vertical or inclined.

Furthermore, this theoretical value of the angle β thus determined is independent of the surface condition of the wall, as is shown by the results of the tests performed "at rest" and also by the experiments relating to the deformation of a retained material when the wall undergoes a small movement.

A.1.7. Remarks

a) The foregoing theoretical propositions have all been verified and confirmed by a large number of experiments performed with granular materials differing widely from one another in their physical and mechanical properties.

b) The angle of repose α, which corresponds to the lower limit of the angle of internal friction of the granular material, is the one which decides the minimum inclination of the path traversed by the elements of the thrust wedge, i.e., it consequently decides the maximum horizontal component of the thrust. It is indeed the only angle to be considered in seeking to determine the maximum value of the thrust exerted by a granular material upon the wall that retains it.

c) We see that the hypotheses which have been made during the course of this treatment of the subject are confirmed by experiment and are thus fully established as valid.

"Any hypothesis, confirmed by experiment, becomes a theory; based solely on logic, it remains a system."

Claude Bernard

A.1.8. Corollary: Determination of the Value of the Thrust Coefficient

Having obtained the thrust coefficient by solving the equilibrium equations for the thrust wedge (Section A.1.4):

$$k_a = \frac{\operatorname{tg} \beta}{\operatorname{tg} \alpha + \operatorname{tg} (\beta + \alpha)}$$

and the value of tg β from formula (1) in Section A.1.5:

$$\operatorname{tg} \beta = \frac{\operatorname{tg} \alpha + \operatorname{tg}\left(\dfrac{\pi}{4} - \dfrac{\alpha}{2}\right)}{1 + \operatorname{tg} \alpha \operatorname{tg}\left(\dfrac{\pi}{4} - \dfrac{\alpha}{2}\right) + \operatorname{tg}^2 \alpha}$$

it is merely necessary to substitute this value of tg β into the general formula for the thrust coefficient k_a to obtain the value thereof.

Table 42 gives the values of the thrust coefficient for the various values of the minimum angle of internal friction φ_0 and the angle of repose α:

Angle $\alpha \cong \varphi_0$	tg β	k_a
24°	0.7359	0.335
27°	0.7140	0.294
30°	0.6928	0.256
33°	0.6720	0.224
36°	0.6507	0.197
39°	0.6301	0.168
45°	0.5858	0.122

Table 42

Table 42 is of considerable interest in that the values of the thrust coefficient k_a contained in it are practically the same as those obtained in Section A.1.9. dealing with the analytical determination of the thrust coefficient.

A.1.9. Theoretical Determination of the Value of the Thrust Coefficient

As shown in Section A.1.4. the equilibrium of the forces acting on the thrust wedge can be written as follows:

$$P = \frac{\gamma \cdot H^2}{2} \cdot tg\ \beta = Q \sin(\beta + \varphi) + R \sin \alpha$$

$$R \cos \alpha = Q \cos(\beta + \varphi)$$

From these equations the following expression for the thrust coefficient is obtained:

$$k_a = \frac{tg\ \beta}{tg\ \alpha + tg\ (\beta + \varphi)}$$

The problem consists in seeking – as Coulomb did – the value of β for which the thrust coefficient k_a is a maximum, i.e., by putting.

$$\frac{dk_a}{d\beta} = 0$$

The value of β obtained from this – lengthy and laborious – derivation is substituted into the above expression for the thrust coefficient and eventually, having performed all the calculations, we obtain the maximum value of k_a:

$$k_{a\,\text{max}} = \left[\frac{(1 - \sqrt{2}\sin\alpha)\cos\alpha}{\cos 2\alpha}\right]^2 \qquad \text{Formula (2)}$$

This is the final expression for the thrust coefficient k_a, the numerical values of which are given in Table 43, as calculated for various values of the minimum angle of internal friction φ, taken as equal to the angle of repose α:

Angle $\alpha \cong \varphi_0$	k_a (max.)
24°	0.335
27°	0.294
30°	0.256
33°	0.224
36°	0.197
39°	0.168
45°	0.122

Table 43

A.1.10. General Conclusion

The experimental investigations described in Volume I enabled us to establish the value of the dihedral angle β of the thrust wedge and to define the value of the maximum thrust coefficient by evaluation and mathematical interpretation of the results of the experiments.

The theoretical formulation dealt with in the present Appendix confirms – by two different methods, the first of which corresponds to strict observation of the actual phenomena, while the second is based on applying the laws of equilibrium to the thrust wedge – the previously obtained results, both for the value of the dihedral angle:

$$\beta = \frac{\pi}{4} - \frac{\alpha}{3}$$

which is thus authenticated, and for the value of the maximum thrust coefficient, relating to the maximum thrust exerted by a mass of granular material with horizontal top surface, retained by a vertical wall:

$$k_a = \left(\frac{\pi - 2\,\alpha}{\pi + 2\,\alpha}\right)^2$$

It is, in addition, to be noted that, as can be deduced from the equilibrium equations for the thrust wedge, the thrust coefficient would, on the basis of Coulomb's assumption that the thrust acts with zero inclination $(\varphi_0 \cong \alpha = 0)$ become:

$$k_a = \frac{\text{tg } \beta}{\text{tg } (\beta + \varphi)}$$

Substituting

$$\beta = \frac{\pi}{4} - \frac{\varphi}{2}$$

this expression is equivalent to:

$$k_a - \text{tg}^2\left(\frac{\pi}{4} - \frac{\varphi}{2}\right)$$

We thus indeed arrive at the conventional thrust coefficient, but it should be borne in mind that this coefficient is valid only on the assumption of zero inclination of the thrust and that the consequence of applying this unconfirmed hypothesis is that the thrust thus defined is larger than the actual thrust – namely, in the ratio of $\pi/(\pi+2\alpha)$, or $\pi/(\pi+2\varphi)$ – so that, in terms of Coulomb's coefficient, the expression for the thrust coefficient should correctly be written thus:

$$k_a = \text{tg}^2\left(\frac{\pi}{4} - \frac{\varphi}{2}\right)\cdot\frac{\pi}{\pi + 2\,\varphi}$$

which is identical with the thrust coefficient of the universal formula previously established in Volume I:

$$k_a = \left(\frac{\pi - 2\,\alpha}{\pi + 2\,\alpha}\right)^2$$

or (see A. 1. 1.):

$$k_a = \left(\frac{\pi - 2\,\varphi}{\pi + 2\,\varphi}\right)^2$$

A.2. Theoretical Justification of the Thrust Coefficient of Retained Material with Inclined Top Surface

It has been shown experimentally (Vol. I, pages 41 to 50) that the thrust coefficient associated with a cohesionless granular material with horizontal top surface is equal to:

$$k_a = \left(\frac{\pi - 2\alpha}{\pi + 2\alpha}\right)^2$$

and that in the case of such a material whose top surface is inclined at an angle α' this coefficient is equal to:

$$k_a = \left(\frac{\pi - 2\alpha}{\pi + 2\alpha}\right)^2 \cdot \left(1 \pm \frac{2\alpha'}{\pi}\right)$$

with the positive sign for material sloping away upward, and the negative sign for material sloping away downward (i.e., below the horizontal) from the retaining wall.

The theoretical justification of the first of these coefficients has been given in Section A.1 after the determination of the angle β of the thrust wedge. Justification – likewise theoretical – of the validity of the second coefficient, which relates to material with inclined top surface, will now be presented below.

A.2.1. Introduction

It is known that, in the case of cohesionless granular materials, the cross-sectional shape (i.e., in a plane perpendicular to the retaining wall) of the thrust wedge is triangular (Fig. A9), namely:

- *ABC* for material with upward-sloping top surface;
- *ABD* for material with horizontal top surface;
- *ABE* for material with downward-sloping top surface.

The angle β is henceforth known, and we also know, from geometric considerations, that the centres of gravity g_1, g_2, g_3 of the above-mentioned triangles are all located on a straight line parallel to *AC* and constituting the locus of the centres of gravity of all the thrust wedge cross-sectional triangles. The distance from this locus to *AC* is equal to a third of the altitude of these triangles with respect to the sliding plane *AC*, whatever the value of the angle α' (whose limits are the upper and the lower angle of repose of the material: $\alpha' = \pm\alpha$).

The phenomenon of the thrust developed by the retained material is therefore simply manifested through the line of action thereof, represented by the line *mn* which intersects the vertical at the point *n*, at a third of the height *AB*. This line *mn* itself (as already noted) is at a third of the distance from the line *AC* to the point *B*, irrespective of whether the top surface of the material is horizontal or inclined.

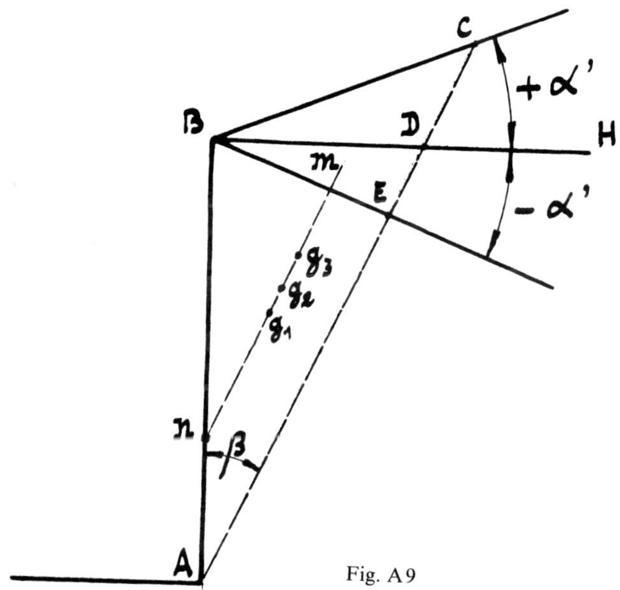

Fig. A 9

A.2.2. Rule of Proportionality

Consider a mass of granular material *ABN* (Fig. A10) whose top surface is inclined at an angle $+\alpha$ corresponding to the angle of repose and which is retained by a vertical wall. This material can be conceived as composed of elements $d\theta$ which sweep its whole area by rotation about the point *B* (top of the wall) through the angular range from the upward slope of the material to the vertical line *AB* of the wall.

It has been shown that for the mass of material *ABH* with horizontal top surface the thrust coefficient is equal to:

$$k_a = \left(\frac{\pi - 2\,\alpha}{\pi + 2\,\alpha}\right)^2$$

The rotation of $d\theta$ between *AB* and *BH* is expressed by a circular function of the coefficient k_a proportionally to the magnitude of the angle θ.

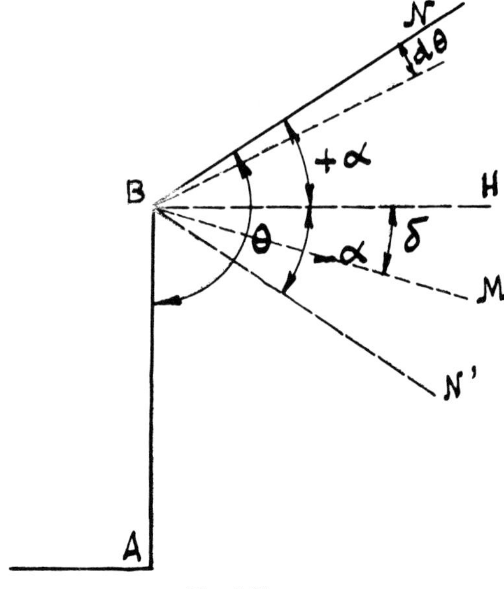

Fig. A 10

In the case of the mass ABN' whose top surface is inclined at an angle $-\alpha$ corresponding to the angle of repose for the material sloping away downward from the wall the phenomenon to be considered is physically possible only from the line AN' onward, with

$$\theta = 90° - \alpha$$

For any intermediate inclination AM, forming an angle δ with the horizontal, the reduction of the thrust coefficient k_a is, by proportionality, equal to:

$$k_a = \frac{|\delta|}{90°}$$

When $\delta = \alpha$, then $k_a = \alpha/90°$, and the thrust coefficient relating to the material ABN' is equal to:

$$k_a = \left(\frac{\pi - 2\,\alpha}{\pi + 2\,\alpha}\right)^2 - \left(\frac{\pi - 2\,\alpha}{\pi + 2\,\alpha}\right)^2 \cdot \frac{\alpha}{90°}$$

$$= \left(\frac{\pi - 2\,\alpha}{\pi + 2\,\alpha}\right)^2 \cdot \left(1 - \frac{\alpha}{90°}\right)$$

or:

$$k_a = \left(\frac{\pi - 2\,\alpha}{\pi + 2\,\alpha}\right)^2 \cdot \left(1 - \frac{2\,\alpha}{\pi}\right)$$

Similarly, when $\delta = +\alpha$, which is the case where the top surface slopes away upward from the wall at an angle equal to the angle of repose, the corresponding thrust coefficient is:

$$k_a = \left(\frac{\pi - 2\,\alpha}{\pi + 2\,\alpha}\right)^2 \cdot \left(1 + \frac{2\,\alpha}{\pi}\right)$$

So we do indeed again arrive at the general coefficient of the universal formula for material with inclined top surface, and the preceding demonstration is also valid in the case of inclined retaining walls.

A.2.3. Utilisation of the Knowledge Concerning the Equilibrium of a Granular Material with Horizontal Top Surface to Confirm the Thrust Coefficient for Material with Inclined Top Surface

A.2.3.1. First Case: Material with Top Surface Inclined at the "Upper" Angle of Repose $+\alpha$ (Sloping Upward from Wall)

The determination of the thrust exerted by a mass of material whose top surface is inclined at an angle $+\alpha$ above the horizontal results from the investigation of the equilibrium of a mass with horizontal top surface which is then conceived as carrying a surcharge of the same granular material as that of which it is itself composed, so as to form the mass with inclined top surface.

This investigation is the direct outcome of applying the fundamental relationship for thrust due to a point load.

A.2.3.1.1. Thrust due to material with horizontal top surface:

It has been shown in Vol. I, page 97, that in the case of a mass of material with horizontal top surface, retained by a vertical wall of height h, the value of the thrust is given with equal validity by the formula:

$$P_1 = \frac{\gamma \cdot h^2}{2} \cdot \left(\frac{\pi - 2\,\alpha}{\pi + 2\,\alpha}\right)^2$$

or:

$$P_1 = \frac{\gamma \cdot h^2}{2} \cdot \mathrm{tg}^4\,\beta \qquad\qquad \beta = \frac{\pi}{4} - \frac{\alpha}{3}$$

A.2.3.1.2. Thrust due to surcharge on material with horizontal top surface; top surface of surcharge is inclined at a given angle:

The surcharge applied to the material with horizontal top surface is:

$$Q = \frac{\gamma \cdot h^2}{2} \cdot tg^2 \, \beta \, tg \, \alpha'$$

and we know that this surcharge produces a thrust equal to:

$$P_2 = \frac{Q}{3} \cdot tg^3 \, \beta$$

A.2.3.1.3. Thrust due to the trapezium ABCC' (Fig. A11) *corresponding to the failure wedge for the material with horizontal top surface carrying a surcharge whose surface is inclined at an angle α (shaded portion in diagram):*

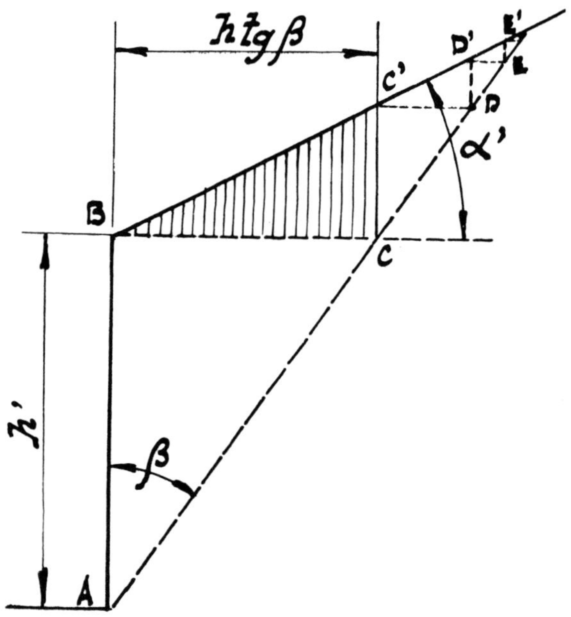

Fig. A 11

The thrust due to the trapezoidal area is the sum of the two thrusts indicated above, the second of these being in this case:

$$P_2 = \frac{\gamma \cdot h^2}{2} \cdot tg^2 \, \beta \, tg \, \alpha' \cdot \frac{tg^3 \, \beta}{3}$$

whence we obtain for the thrust associated with the trapezium $ABCC'$ $(P_1 + P_2)$:

$$P = \frac{\gamma \cdot h^2}{2} \cdot \left[\mathrm{tg}^4\, \beta + \frac{\mathrm{tg}^5\, \beta\ \mathrm{tg}\, \alpha'}{3} \right]$$

or, putting:

$$P = \frac{\gamma \cdot h^2}{2} \cdot k_a$$

we obtain for k_a, which is the thrust coefficient relating to this trapezium $ABCC'$, the following expression:

$$k_a = \mathrm{tg}^4\, \beta + \frac{\mathrm{tg}^5\, \beta\ \mathrm{tg}\, \alpha'}{3}$$

With this value of k_a we can determine (see Section A.2.3.1.3) the thrust due to material with inclined top surface.

A.2.3.1.4. Thrust due to granular material with top surface inclined at "upper" angle of repose $+\alpha$:

According to Fig. A11, the failure wedge can be subdivided into successive trapezia $ABCC'$, $CC'DD'$, $DD'EE'$, etc. with diminishing areas, whose parallel bases, terminating at the general plane of sliding of the thrust wedge, have the following successive values:

$$CC' = h\ \mathrm{tg}\, \beta\ \mathrm{tg}\, \alpha$$

$$DD' = h\ \mathrm{tg}^2\, \beta\ \mathrm{tg}^2\, \alpha$$

$$EE' = h\ \mathrm{tg}^3\, \beta\ \mathrm{tg}^3\, \alpha$$

$$\cdots\cdots \qquad \cdots\cdots\cdots\cdots$$

According to the foregoing formula, the general expression for the thrust due to any particular trapezium whose major base is h_n would be of the following form:

$$P_n = \frac{\gamma \cdot h_n^2}{2} \cdot \left[\mathrm{tg}^4\, \beta + \frac{\mathrm{tg}^5\, \beta\ \mathrm{tg}\, \alpha}{3} \right] \mathrm{tg}^n\, \beta\ \mathrm{tg}^n\, \alpha$$

Hence it follows that the thrust due to the mass of material with inclined top surface – or the sum of the thrusts associated with the successive trapezia whose major bases are AB, CC', DD', EE', etc. – is expressed by:

$$\frac{\gamma \cdot h^2}{2} \cdot \left[\text{tg}^4\, \beta + \frac{\text{tg}^5\, \beta\ \text{tg}\ \alpha}{3} \right] [1 + \text{tg}^2\, \beta\ \text{tg}\ \alpha + \text{tg}^4\, \beta\ \text{tg}^4\, \alpha + \text{tg}^6\, \beta\ \text{tg}^6\, \alpha \ldots]$$

This is equivalent to stating that the thrust coefficient associated with the mass of material with top surface inclined at the "upper" angle of repose $+\alpha$ is:

$$k_a = \left[\text{tg}^4\, \beta + \frac{\text{tg}^5\, \beta\ \text{tg}\ \alpha}{3} \right] [1 + \text{tg}^2\, \beta\ \text{tg}^2\, \alpha + \text{tg}^4\, \beta\ \text{tg}^4\, \alpha + \text{tg}^6\, \beta\ \text{tg}^6\, \alpha + \ldots]$$

$$= \left[\text{tg}^4\, \beta + \frac{\text{tg}^5\, \beta\ \text{tg}\ \alpha}{3} \right] \cdot \frac{1}{1 - \text{tg}^2\, \beta\ \text{tg}^2\, \alpha}$$

and it can be shown that, in the limiting case, this coefficient is equivalent to:

$$k_a = \left(\frac{\pi - 2\,\alpha}{\pi + 2\,\alpha} \right)^2 \cdot \left(1 + \frac{2\,\alpha}{\pi} \right)$$

This result therefore indeed confirms the validity of the expression obtained, by interpretation of the test results, for the thrust coefficient associated with retained granular material whose top surface is inclined at the "upper" angle of repose $+\alpha$.

A.2.3.1.5. Numerical application providing confirmation of the validity of the expression for the thrust exerted by a granular material whose top surface is inclined at an angle $+\alpha$:

In this example we wish to determine the thrust exerted by material of density γ whose top surface is inclined at $30°$ and which is retained by a 10.00 m high wall. The minimum angle of internal friction φ_0 can be taken as equal to the angle of repose.

According to Fig. A11 the thrust wedge is subdivided into successive trapezia $ABCC'$, $CC'DD'$, $DD'EE'$, $EE'FF'$, etc. whose major bases, expressed in terms of the angle of the thrust wedge

$$\beta = \frac{\pi}{4} - \frac{30°}{3} = 35°$$

are respectively equal to:

$$
\begin{aligned}
AB &= & 10.00 \\
CC' &= 10.00 \ \text{tg} \ 35° \ \text{tg} \ 30° = & 4.04 \\
DD' &= 4.04 \ \text{tg} \ 35° \ \text{tg} \ 30° = & 1.632 \\
EE' &= 1.632 \ \text{tg} \ 35° \ \text{tg} \ 30° = & 0.66 \\
\dots\ & 0.66 \ \text{tg} \ 35° \ \text{tg} \ 30° = & 0.266 \\
\dots\ & 0.266 \ \text{tg} \ 35° \ \text{tg} \ 30° = & 0.107 \\
\dots\ & 0.107 \ \text{tg} \ 35° \ \text{tg} \ 30° = & 0.043
\end{aligned}
$$

According to the formula in Section A.2.3.1.3. the thrust coefficient corresponding to any particular trapezium of the subdivision of the thrust wedge in accordance with Section A.2.3.1.4. is:

$$
k_a = \text{tg}^4 \ 35° + \frac{\text{tg}^5 \ 35° \ \text{tg} \ 30°}{3} = 0{,}272
$$

The thrusts due to the successive trapezia are therefore:

Total thrust:

$$
\begin{aligned}
ABCD \quad & 1/2 \ \gamma \cdot \overline{10.00}^2 \ \times 0.272 = & 13.613 \ \gamma \\
CC'DD' \quad & 1/2 \ \gamma \cdot \overline{4.04}^2 \ \times 0.272 = & 2.220 \ \gamma \\
DD'EE' \quad & 1/2 \ \gamma \cdot \overline{1.632}^2 \times 0.272 = & 0.362 \ \gamma \\
EE''FF \quad & 1/2 \ \gamma \cdot \overline{0.66}^2 \ \times 0.272 = & 0.059 \ \gamma \\
FF'GG' \quad & 1/2 \ \gamma \cdot \overline{0.266}^2 \times 0.272 = & 0.0096 \ \gamma \\
GG'HH' \quad & 1/2 \ \gamma \cdot \overline{0.107}^2 \times 0.272 = & 0.0016 \ \gamma \\
HH'II' \quad & 1/2 \ \gamma \cdot \overline{0.043}^2 \times 0.272 = & 0.0002 \ \gamma \\
\hline
& & 16.2654 \ \gamma
\end{aligned}
$$

We thus see that this result indeed confirms, with satisfactory approximation, the value of the thrust due to the material with inclined top surface and calculated by means of the universal formula:

$$
P = \frac{\gamma \cdot \overline{10.00}^2}{2} \cdot \left(\frac{\pi - 2 \times 30°}{\pi + 2 \times 30°} \right)^2 \cdot \left(1 + \frac{2 \times 30°}{180°} \right)
$$

$$
= 16.6 \ \gamma
$$

A.2.3.2. Second Case: Material with Top Surface Inclined at the "Lower" Angle of Repose $-\alpha$ (Sloping Downward from Wall)

Analytical confirmation of the formula for the thrust associated with a mass of material whose top surface is inclined at the "lower" angle of repose $-\alpha$ is obtained as follows (as was done previously, in 1965, see Vol. I, p. 57, before the fundamental relationship for thrust due to a point load was established in 1969):

The thrust exerted by material whose top surface is inclined at the "lower" angle of repose can be conceived as the thrust due to a mass whose top surface is horizontal and which has a height BD (Fig. A12) corresponding to the point C where the sliding plane of the thrust wedge – forming an angle $\beta = \pi/4 - \alpha/3$ with the vertical – intersects the inclined top surface of the retained material, to which must be added the thrust due to a triangular surcharge (ADC).

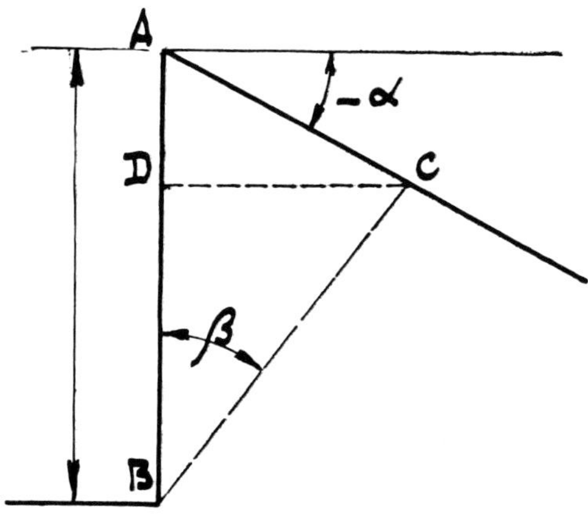

Fig. A 12

The thrusts in question are to be calculated as follows:

$$DC = AD \cot g\,\alpha \quad \text{and} \quad DC = BD \operatorname{tg}\alpha$$

whence we obtain:

$$AD = h \cdot \frac{\operatorname{tg}\beta}{\cot g\,\alpha + \operatorname{tg}\beta}$$

$$BD = h\left(1 - \frac{\operatorname{tg}\beta}{\cot g\,\alpha + \operatorname{tg}\beta}\right)$$

A.2.3.2.1. Thrust due to material with horizontal top surface (material within BCD):

We know from page 45, Vol. 1, that this thrust is equal to:

$$P_1 = \frac{\gamma \cdot \overline{BD}^2}{2} \cdot \left(\frac{\pi - 2\alpha}{\pi + 2\alpha}\right)^2$$

or:

$$P_1 = \frac{\gamma \cdot h^2}{2} \cdot \left(1 - \frac{\operatorname{tg}\beta}{\operatorname{cotg}\alpha + \operatorname{tg}\beta}\right)^2$$

A.2.3.2.2.   Thrust due to surcharge on material with horizontal top surface (material within BCD) exerted by the triangular surcharge ADC which completes the retained mass:

The triangular surcharge (triangle *ADC* in Fig. A 12), applied to the material with horizontal top surface *CD* already referred to, is:

$$Q - \gamma \cdot \frac{\overline{AD} \times \overline{DC}}{2} = \frac{\gamma \cdot h^2}{2} \cdot \frac{\operatorname{tg}^2\beta \operatorname{cotg}\alpha}{(\operatorname{cotg}\alpha + \operatorname{tg}\beta)^2}$$

We know that such a surcharge diminishing linearly from the retaining wall to the rear end *C* of the thrust wedge produces a thrust equal to:

$$P_2 = \frac{2Q}{3} \cdot \operatorname{tg}^3\beta$$

or:

$$P_2 = \frac{\gamma \cdot h^2}{2} \cdot \frac{\operatorname{tg}^5\beta \operatorname{cotg}\alpha}{(\operatorname{cotg}\alpha + \operatorname{tg}\beta)^2}$$

A.2.3.3. Thrust Due to Granular Material with Top Surface Inclined at "Lower" Angle of Repose $-\alpha$:

This thrust is the sum of the two thrusts envisaged above, namely, $P = P_1 + P_2$ or:

$$P = \gamma \cdot h^2 \left[\frac{1}{2}\left(1 - \frac{\operatorname{tg}\beta}{\operatorname{cotg}\alpha + \operatorname{tg}\beta}\right)^2 \cdot \left(\frac{\pi - 2\alpha}{\pi + 2\alpha}\right)^2 + \frac{1}{3} \cdot \frac{\operatorname{tg}^5\beta \operatorname{cotg}\alpha}{(\operatorname{cotg}\alpha + \operatorname{tg}\beta)^2}\right]$$

We can verify that this result is equivalent to the value determined for the thrust by interpretation of the test results (page 50, Vol. 1):

$$P = \frac{\gamma \cdot h^2}{2} \cdot tg^4 \left(\frac{\pi}{4} - \frac{\alpha}{3} \right) \cdot \left(1 - \frac{2\,\alpha}{\pi} \right)$$

A.2.3.3.1. Numerical application providing confirmation of the validity of the expression for the thrust exerted by a granular material whose top surface is inclined at an angle $-\alpha$:

In this example we wish to determine the thrust exerted by material of density γ whose top surface is inclined at 30° and which is retained by a 10.00 m high wall. The minimum angle of internal friction φ_0 can be taken as equal to the angle of repose.

The thrust wedge angle is:

$$\beta = \frac{\pi}{4} - \frac{30°}{3} = 35°$$

According to the formula in Section A.2.3.3. the thrust is:

$$P = \gamma \cdot \overline{10.00}^2 \left[\frac{1}{2} \left(1 - \frac{tg\,35°}{cotg\,30° + tg\,35°} \right) \cdot \left(\frac{\pi - 2 \times 30°}{\pi + 2 \times 30°} \right)^2 \right.$$

$$\left. + \frac{1}{3} \cdot \frac{tg^5\,35°\;cotg\,30°}{(cotg\,30° + tg\,35°)^2} \right]$$

and we can verify that this result is the same as that obtained by applying the formula obtained by interpretation of the test results:

$$= 8\,\gamma$$

$$P = \frac{\gamma \cdot \overline{10.00}^2}{2} \cdot tg^4 \left(\frac{\pi}{4} - \frac{30°}{3} \right) \cdot \left(1 - \frac{2 \times 30°}{180°} \right)$$

$$= 8\,\gamma$$

A.3. Theoretical Justification of the Thrust Coefficient of Retained Material Carrying a Surcharge

A.3.1. Uniformly Distributed Surcharge

The thrust due to the surcharge is relatively complex. Its theoretical justification and explanation will be based on the experimental investigation comprising numerous tests which have revealed a characteristic fact that distinguishes the case of non-surcharged granular materials from that of such materials carrying surcharge.

In the case of non-surcharged materials retained by a wall the thrust diagram is known to be linear (Fig. A13) with the maximum value of the active pressure acting at the base of the wall.

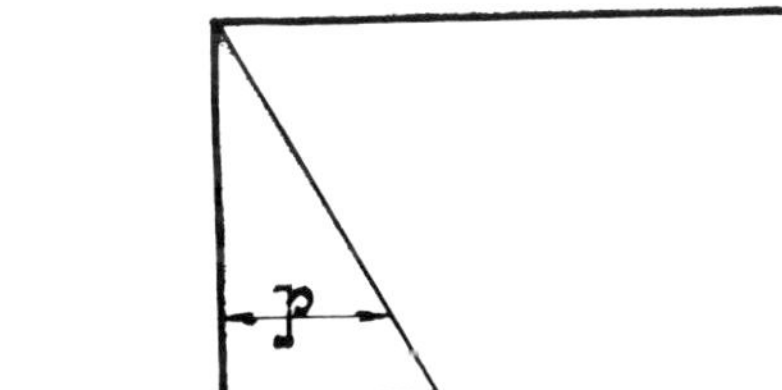

Fig. A13

It has been shown by experiment that in the case of a retained material carrying a uniformly distributed surcharge the thrust diagram (active pressure distribution) due to surcharge is again linear, but now the active pressure is zero at the base of the wall and has its maximum value at the top (Fig. A14).

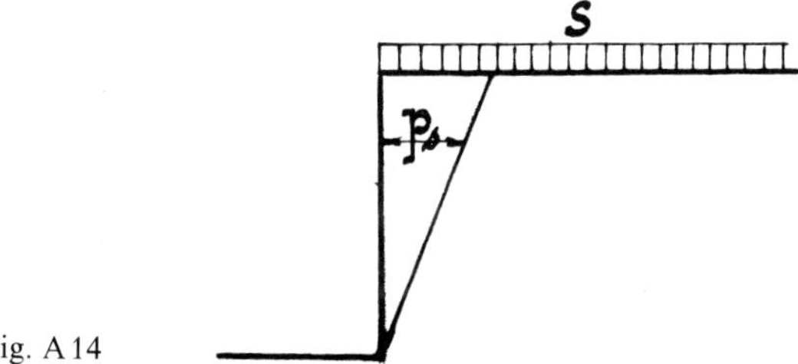

Fig. A14

So there is an inversion of the phenomenon of thrust, according to whether we are concerned with the internal equilibrium of the actual material subjected to the action of its own weight, with due regard to friction, or whether we are concerned with external imposed loading.

It follows that the direction of the reaction of the material in opposition to the action of the surcharge justifies the inversion of the diagrams in Figs. A 13 and A 14, as has indeed been verified experimentally (see Vol. I, pages 92, 97: effect of surcharge on the retained material), and that the diagram in Fig. A 14 represents the "reaction effect opposed to the surcharge" and is inclined at an angle $\beta = \pi/4 - \alpha/3$, with respect to the vertical, since we have seen (Section A 1) that this angle is solely a function of the angle α and independent of the density γ of the material, and that it is the same whether or not there is a surcharge on the material (see Vol. I, page 108 to 112: experiments relating to uniformly distributed surcharge).

On the other hand, the loading to which the grains of the retained material are subjected by the action of the surcharge produces a compression of the grains. The effect of such compression is similar to an increase γ' in the density of the granular material.

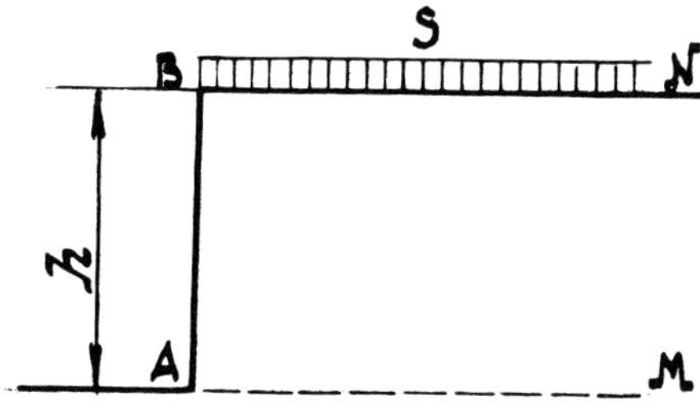

Fig. A 15

According to Fig. A 15, between the horizontal planes AM and BN (whose distance apart is equal to the height h of the wall) the surcharge S per unit area produces a virtual increase in density equal to the vertical uniform average pressure throughout the mass of material, this increase in density being:

$$\gamma' = \frac{S}{h}$$

We may thus consider the corresponding thrust P_s (Fig. A 14) in terms of this density γ':

$$P_s = \frac{\gamma' h^2}{2} \cdot \left(\frac{\pi - 2\alpha}{\pi + 2\alpha}\right)^2$$

or:

$$P_s = \frac{S}{h} \cdot \frac{h^2}{2} \cdot \left(\frac{\pi - 2\alpha}{\pi + 2\alpha}\right)^2$$

The total thrust due to the surcharged material is then:

$$P_T = P + P_s = \frac{\gamma \cdot h^2}{2} \cdot \left(\frac{\pi - 2\,\alpha}{\pi + 2\,\alpha}\right)^2 + \frac{S}{h} \cdot \frac{h^2}{2} \cdot \left(\frac{\pi - 2\,\alpha}{\pi + 2\,\alpha}\right)^2$$

or:

$$P_T = \frac{\gamma \cdot h^2}{2} \cdot \left(\frac{\pi - 2\,\alpha}{\pi + 2\,\alpha}\right)^2 \cdot \left(1 + \frac{S}{\gamma \cdot h}\right)$$

Hence we do indeed arrive again at the expression for the thrust exerted by a surcharged material as defined in the universal formula.

Besides, this expression was subsequently confirmed by application of the fundamental relationship for thrust due to a point load (Vol. I, page 133) and furthermore by the experiments relating to the deformations of a mass of material (Section 11 of Vol. II). Hence it follows that this is a generally-valid expression, applicable to surcharged granular materials, whether they develop active pressure or whether they develop passive resistance, up to the stage of plastic equilibrium thereof, corresponding to their deformation due to displacement of the retaining wall.

A.3.2. Locally Surcharged Material Retained by a Wall

The foregoing justification of the formula for the thrust due to a uniformly distributed surcharge involves, ipso facto, the justification of the formula for the thrust due to a local load Q applied at a distance x from the rear boundary of the thrust wedge (Fig. A16):

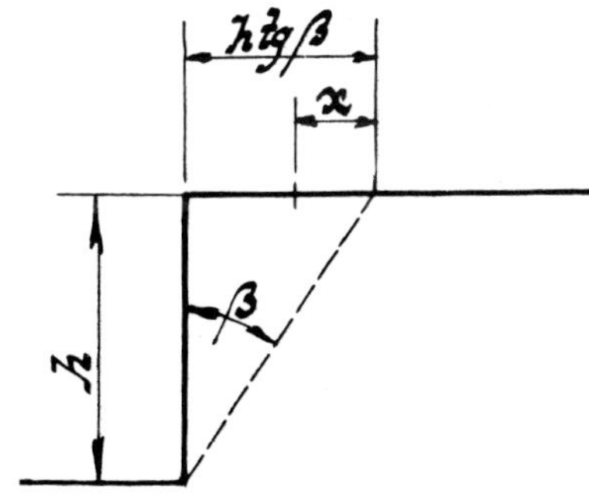

Fig. A16

$$P_q = Q \cdot \operatorname{tg}^2 \beta \cdot \frac{x}{h}$$

where

$$\beta = \frac{\pi}{4} - \frac{\alpha}{3}$$

In fact, it need only be noted that the thrust due to the uniformly distributed surcharge, written in the form:

$$P_s = \frac{S}{h} \cdot \frac{h^2}{2} \cdot \operatorname{tg}^4 \beta$$

is, because of the equivalence of:

$$\left(\frac{\pi - 2\,\alpha}{\pi + 2\,\alpha} \right)^2$$

and

$$\operatorname{tg}^4 \left(\frac{\pi}{4} - \frac{\alpha}{3} \right)$$

the integral of the active pressure p_x due to the elementary load $\dfrac{S}{h}\,dx$:

$$P_q = \int_0^{h\,\operatorname{tg}\beta} \frac{S}{h} \cdot \operatorname{tg}^2 \beta \cdot x\,dx$$

or:

$$P_q = \frac{S}{h} \cdot \frac{h^2}{2} \cdot \operatorname{tg}^4 \beta$$

Thus we have a theoretical justification of the experimentally established fundamental relationship for the thrust due to a point load.

Important Remark

It must once again be emphasised that the value of the minimum angle of internal friction φ_0 of a granular material tends, in the limiting case, to the value of the angle of repose α, and that this fact must be seen as the reason why the angle of the thrust wedge – which, in the limiting case, corresponds to the failure wedge of the retained material – remains equal to $\beta = \pi/4 - \alpha/3$, whether or not the material carries a surcharge.

This emerges from the laws of strength of materials, for if the minimum angle of internal friction were to undergo some minor variations under the effect of the compression produced by the surcharge acting on the top surface of the retained material, the dihedral angle of the thrust wedge $\beta = \pi/4 - \alpha/3$ would nevertheless remain entirely unchanged, as failure always occurs in accordance with the minimum angle of internal friction, this angle then having become equal to the angle of repose, as has been demonstrated experimentally (Vol. II, Section 20).

SYNOPSIS

Our experimental results – relating to the investigation of the equilibrium of granular materials of any kind, with horizontal or inclined top surface and with or without surcharge, retained by a vertical or an inclined wall, and all embodied in simple formulas (with no more than the essential mathematics for a proper understanding of the subject) – have subsequently been explained and justified theoretically.

Henceforth these results are characterised by a generalisation of those formulas in their rigorous interdependence, without any internal contradiction, whatever the type of loading conditions to which the materials under consideration are subjected, i.e., whether exerting active pressure or developing passive resistance.

This absolute generalisation therefore constitutes the basis of a fundamental relationship interpreting the whole of the natural phenomena investigated and reported.

Errata

Please correct the following items in Volume I Retaining Walls:

1. Page 26, 22nd line: Chapter 13 instead of Chapter 12
2. Page 222, 9th line: $h = l \cos i$ instead of $h = l \sin i$
3.   18th line: $\cos^2 i$ instead of $\sin^2 i$
4. Page 223, 17th line: $\cos^2 i$ instead of $\sin^2 i$

References

(1)   Reimbert, M. & A.:
Retaining Walls, Anchorages and Sheet Piling Vol. I.
Trans Tech Publications, Clausthal 1974

(2)   Reimbert, M. & A.:
Annales No. 211–212, July/August 1965, Institut Technique du Bâtiment et Travaux Publics à Paris

(3)   Reimbert, M. & A.:
Proceedings of the 5th European Congress on Soil Mechanics, Madrid, Vol. II (1972)

(4)   Rowe, P. W.:
Passive Earth Pressure Measurements.
Geotechnique, Vol. 25 (1965), No. 1

(5)   Kérisel, J.:
The Language of Models. 5th European Congress on Soil Mechanics, Madrid, Vol. II (1972)

(6)   Caquot, A.:
Equilibre des massifs a frottement interne (Equilibrium of Materials with Internal Friction). Publ. Gauthires-Villars, Paris, 1934

(7)   Buisson, M.:
Essais de geotechnique. Caracteristiques physiques et mecaniques des sols (Geotechnical Tests. Physical and Mechanical Characteristics of Soils).
Editions Dunod, 1942

(8)   Terzaghi, K. & Peck, R. B.:
Soil Mechanics in Engineering Practice. John Wiley, New York, 1948

(9)   Mayer, A:
Sols et Fondations. Ed. Azmand Colin, Paris 1942

(10) Caquot, A. & Kérisel, J.:
Traité de Mecanique des Sols, 1966

(11) Résal, J.:
Cours de l'Ecole des Ponts et Chaussées de PARIS – Librairie Polytechnique Ch. BERANGER, Editeur – PARIS –
1.: Poussée des terres, stabilité des Murs de Soutènement – 1903,
2.: Poussée des terres, théorie des terres cohérentes, applications, tables numériques – 1910.

(12) Kérisel, J.:
Memoirs de Coulomb. 8th International Congress on Soil and Foundation Mechanics, Moscow 1973

Series on Rock and Soil Mechanics
Vol. 1 (1971/74) No. 3

FOUNDATION INSTRUMENTATION

By **Dr. Thomas H. Hanna,** Professor of Civil and Structural Engineering,
University of Sheffield, England

1973, 372 pages, 251 figures, 520 references, price: US $ 35.00 hard cover

International Standard Book Number: ISBN 0-87849-006-x
Library of Congress Catalog Card Number: LC 72-90015

Contents

1. Introduction
2. Load Measurement
3. Pore Water Pressure Measurement
4. Earth Pressure Measurement
5. Measurement of Ground Movements
6. Data from Instrumented Foundations
7. The Recording and Processing of Field Data
8. Instrumentation of Laboratory Scale Foundations
9. Appendix

"The book represents a fine help and a welcome treasury of methods and devices for every civil and structural engineer concerned with the design and construction of civil engineering works, since the ground always affects the stability and performance of these structures. It can be recommended warmly to students and civil engineers in the field of design, construction and research."

Applied Mechanics Reviews

"This most interesting book includes a very large number of references and a list of instrument suppliers. It will probably become one of the most widely used tools for soil and foundation engineers who understand the need for performance evaluation."

Canadian Geotechnical Journal

"The book can obviously be recommended to all people dealing with foundations, earth and rockfill dams, tunnels, and soil mechanics in general."

Water Power